Bioavailability of Metals in Terrestrial Ecosystems: Importance of Partitioning for Bioavailability to Invertebrates, Microbes, and Plants

Other titles from the Society of Environmental Toxicology and Chemistry (SETAC):

Development of Methods for Effects-Driven Cumulative Effects Assessment Using Fish Populations: Moose River Project
Munkittrick, McMaster, Van der Kraak, Portt, Gibbons, Farwell, Gray
2000

Evaluation of Persistence and Long-Range Transport of Organic Chemicals in the Environment
Klečka, Boethling, Franklin, Grady, Graham, Howard, Kannan, Larson, Mackay, Muir, van de Meent, editors
2000

Ecotoxicology and Risk Assessment for Wetlands
Lewis, Mayer, Powell, Nelson, Klaine, Henry, Dickson, editors
1999

Advances in Earthworm Ecotoxicology
Sheppard, Bembridge, Holmstrup, Posthuma, editors
1998

Ecotoxicological Risk Assessment of the Chlorinated Organic Chemicals
Carey, Cook Geisy, Hodson, Muir, Owens, Solomon, editors
1998

Atmospheric Deposition of Contaminants to the Great Lakes and Coastal Waters
Baker, editor
1997

Reassessment of Metals Criteria for Aquatic Life Protection: Priorities for Research and Implementation
Bergman and Dorward-King, editors
1997

For information about any SETAC publication, including SETAC's international journal, *Environmental Toxicology and Chemistry*, contact the SETAC Office nearest you:

1010 North 12th Avenue
Pensacola, Florida, USA 32501-3367
T 850 469 1500
F 850 469 9778
E setac@setac.org

Avenue de la Toison d'Or 67
B-1060 Brussels, Belgium
T 32 2 772 72 81
F 32 2 770 53 86
E setac@setaceu.org

www.setac.org

Environmental Quality Through Science®

Bioavailability of Metals in Terrestrial Ecosystems: Importance of Partitioning for Bioavailability to Invertebrates, Microbes, and Plants

Edited by

Herbert E. Allen
Department of Civil and Environmental Engineering
University of Delaware
Newark, Delaware, USA

With contributions by

Stephen P. McGrath
IACR – Rothamsted
Harpenden, Hertfordshire, United Kingdom

Michael J. McLaughlin
CSIRO Australia
Adelaide, South Australia, Australia

Willie J.G.M. Peijnenburg
National Institute of Public Health and the Environment
Bilthoven, Netherlands

Sébastien Sauvé
University of Montréal
Montréal, Québec, Canada

Metals and the Environment Series Editor
Chris Lee
International Copper Association
New York, New York, USA

Current Coordinating Editor of SETAC Books
Andrew Green
International Lead Zinc Research Organization
Department of Environment and Health
Durham, North Carolina, USA

Publication sponsored by the Society of Environmental Toxicology and Chemistry (SETAC)

Cover by Michael Kenney Graphic Design and Advertising
Copyediting and typesetting by Wordsmiths Unlimited
Indexing by IRIS

Library of Congress Cataloging-in-Publication Data

Bioavailability of metals in terrestrial ecosystems: importance of partitioning for bioavailability to invertebrates, microbes, and plants / edited by Herbert E. Allen : with contributions by Stephen McGrath...[et al.].
p. cm.
This publication is a result of revised, recirculated and finalized material of a conference.
Includes bibliographical references.
ISBN 1-880611-46-5 (alk. paper)
1. Metals--Environmental aspects. 2. Bioavailability--Congress. I. Allen, Herbert E. (Herbert Ellis), 1939–

QH545.M45 B56 2001
577'.1475--dc21

2001054170

This publication is printed on recycled paper using soy ink.
SETAC Press is an imprint of the Society of Environmental Toxicology and Chemistry.
No claim is made to original U.S. Government works.

International Standard Book Number 1-880611-46-5
Printed in the United States of America
09 08 07 06 05 04 03 02 10 9 8 7 6 5 4 3 2

∞ The paper used in this publication meets the minimum requirements of the American National Standard for Information Sciences—Permanence of Paper for Printed Library Materials, ANSI Z39.48-1984.

Reference Listing: Allen HE. 2002. Bioavailability of metals in terrestrial ecosystems: Importance of partitioning for bioavailability to invertebrates, microbes, and plants. Pensacola FL: Society of Environmental Toxicology and Chemistry (SETAC). 176 p.

Metals and the Environment Series

This book is the first in the Metals and the Environment (MATE) Series, which originated during discussions of the Ecotoxicity Technical Advisory Panel (ETAP), a group sponsored by metals research organizations and committed to furthering research and understanding of metals in the environment.

The MATE Series is dedicated to the memory of Dr. Christopher M. Lee, who was a driving force for the creation of the ETAP and the creator of the concept for the MATE Series of publications. His leadership and vision on environmental issues will be missed, but his legacy provides a firm foundation for others to build upon.

SETAC Publications

The publication of books by the Society of Environmental Toxicology and Chemistry (SETAC) provides in-depth reviews and critical appraisals on scientific subjects relevant to understanding the impacts of chemicals and technology on the environment. The books explore topics reviewed and recommended by the Publications Advisory Council and approved by the SETAC Board of Directors for their importance, timeliness, and contribution to multidisciplinary approaches to solving environmental problems. The diversity and breadth of subjects covered in the series reflect the wide range of disciplines encompassed by environmental toxicology, environmental chemistry, and hazard and risk assessment. These volumes attempt to present the reader with authoritative coverage of the literature, as well as paradigms, methodologies, and controversies; research needs; and new developments specific to the featured topics. The books are generally peer reviewed for SETAC by acknowledged experts.

SETAC Publications, which include Technical Issue Papers (TIPs), workshop summaries, newsletter (*SETAC Globe*), and journal (*Environmental Toxicology and Chemistry*), are useful to environmental scientists in research, research management, chemical manufacturing and regulation, risk assessment, and education, as well as to students considering or preparing for careers in these areas. The publications provide information for keeping abreast of recent developments in familiar subject areas and for rapid introduction to principles and approaches in new subject areas.

SETAC would like to recognize the past SETAC Special Publication Series editors:

Table of Contents

List of Figures

List of Tables

Acknowledgments

The contributors and I wish to express our thanks to the International Copper Association (ICA), the International Lead Zinc Research Organization (ILZRO), and the Nickel Producers Environmental Research Association (NiPERA), the sponsors of the Ecotoxicity Technical Advisory Panel (ETAP), for making it possible for us to prepare this document. We are all indebted to the governmental and industrial organizations that have sponsored our research. They have been responsible for the advances in knowledge that have occurred in our laboratories. We have pointed the way for further research at our laboratories and those of the many others who are endeavoring to understand the bioavailability of trace metals in soils. We are appreciative of the careful review by E. Smolders and C.A.M. van Gestel. Finally, we wish to thank Dana M. Crumety, Melissa Winter, the SETAC Office, and Mimi Meredith of Wordsmiths Unlimited for their endeavors in preparing the final manuscript for publication.

— Herbert E. Allen

About the Editor

Herbert E. Allen, PhD, is Professor of Environmental Engineering at the University of Delaware. Previously, he served on the faculty of Drexel University in Philadelphia and the Illinois Institute of Technology in Chicago. Dr. Allen received his PhD in Environmental Chemistry from the University of Michigan.

Dr. Allen's principal areas of research are methods to predict metal availability in both aquatic and terrestrial systems for the development of environmental quality criteria. He has authored more than 160 technical publications, has edited 7 books, and has prepared numerous reports.

Dr. Allen is active in professional organizations, particularly the Society of Environmental Toxicology and Chemistry (SETAC) and the American Chemical Society's (ACS) Division of Environmental Chemistry. He has served as a consultant to a number of industrial companies, to both U.S. and foreign government agencies, and to the World Health Organization. He was a member of the Ecotoxicity Technical Advisory Panel (ETAP) for the International Copper Association (ICA), the International Lead Zinc Research Organization (ILZRO), and the Nickel Producers Environmental Research Association (NiPERA) from 1995 through 1999.

Preface

This publication originated during discussions of the Ecotoxicity Technical Advisory Panel (ETAP), a group sponsored by the International Copper Association (ICA), the International Lead Zinc Research Organization (ILZRO), and the Nickel Producers Environmental Research Association (NiPERA). These discussions culminated in the establishment of the task force that prepared this book. The objectives were to define the state of the art of measurement and prediction of bioavailability of trace metals in terrestrial ecosystems and the means by which these can be incorporated into the regulatory framework.

Draft versions of the chapters were prepared and circulated among the members of the task group. Sébastien Sauvé prepared the draft of Chapter 2 "Speciation of Metals in Soils," Michael J. McLaughlin prepared the draft of Chapter 3 "Bioavailability of Metals to Terrestrial Plants," Stephen P. McGrath prepared the draft of Chapter 4 "Bioavailability of Metals to Soil Microbes," and Willie J.G.M. Peijnenburg prepared the draft of Chapter 5 "Bioavailability of Metals to Soil Invertebrates." We discussed the drafts in Vienna, Austria, 10 to 11 July 1999, immediately preceding the Fifth International Conference on the Biogeochemistry of Trace Elements. The materials were revised, recirculated, and finalized for this publication.

Chapter 1

Terrestrial Ecosystems: An Overview

Herbert E. Allen
University of Delaware

Criteria and standards for metals in soils are important in hazard and risk assessment. Such values have been used as limits for the amount of a metal that is deemed to be acceptable in a soil, for example, for the application of sewage sludges, or in establishing policies for material use, such as the acceptability of Cu and Zn roofing materials. They are also used in remediation to differentiate the soil requiring treatment from the soil not requiring treatment. The criteria are intended to protect desired uses of the soil and include protection of groundwater, uptake of metals by plants and soil-dwelling organisms, the food chain, and humans and animals that directly ingest soil.

Criteria and standards for metals in soils usually are based on total metal concentration as determined by acid digestion. The criteria usually are based on the lowest concentration that has been reported to produce the undesired effect. Criteria based on total concentrations are generally conservative. Criteria that incorporate bioavailability of metals would be site specific and predictive of actual effects. A key factor controlling bioavailability is the strength of binding of the metal by the soil. The partitioning of metal to the soil reduces the availability of the metal for mobilization and uptake by plants, animals, and microbes. The strength of binding varies markedly across soils as a consequence of differences in soil properties. Among the key properties of soils controlling the partitioning process are

- soil solution pH,
- dissolved organic matter (DOM) and Ca, and
- solid-phase metal oxide and organic matter content.

The partitioning may be modified in the rhizosphere and within the digestive tract of soil-dwelling organisms. Partitioning may also change over time because of natural or anthropogenic processes, for example, acidification, salinization, or organic matter mineralization. Soil amendments for the purpose of remediation are designed to modify metal availability. Bioavailability of metals to plants has been correlated to free metal ion activity and to the diffusion of metals in the soil pore water. Toxicity of metals to soil microorganisms also has been related to free metal ion activity in pore water. Invertebrates also are exposed to metals when organic matter in the soil is digested in their guts. However, in some cases, bioavailability

Bioavailability of Metals in Terrestrial Ecosystems: Importance of Partitioning for Bioavailability to Invertebrates, Microbes, and Plants.
Herbert E. Allen, editor. © 2002 Society of Environmental Toxicology and Chemistry (SETAC). ISBN 1-880611-46-5

has been found to correlate better with total metal than with free metal ion concentration.

Soils are heterogeneous materials in which trace metals (e.g., As, Cd, Co, Cr, Cu, Hg, Ni, Pb, Se, and Zn) can exist in a variety of forms. With the exception of As, Cr, and Se, the metals are cationic. The soil chemistry of oxyanions of As, Cr, and Se will not be considered in this review. Differences in binding mechanisms and strength of binding, along with differences in the chemistry of the solution phase into which the metals are solubilized, lead to the observed variations of the partitioning of these metals and to the differences in the availability of trace metals from soil to organisms. Therefore, total metal concentrations in the soil are not predictive of the potential risk that a metal concentration in soil presents. For some organisms, the metal concentration in soil solution may be predictive of metal availability. Organisms may modify the environment of soil particles such that the partitioning of the metals is altered. For example, such factors as the pH of the gut of a soil-dwelling invertebrate may play an important role in controlling the availability of metals. It is the ability of the metal to partition from the solid phase to a soluble phase to a reaction with an important receptor site that leads to risk to the organism.

The chemical interactions important in the partitioning of metal bound to soil surfaces (MeSur), to the solution phase, and then to a plant are shown in Figure 1-1, while the interactions important to the transfer to invertebrates and microbes are shown in Figures 1-2 and 1-3, respectively. Important factors controlling partitioning of metals between soil and soil solution include pH, inorganic ligands (L_{inorg}, e.g., HCO_3^- and Cl^-), organic ligands (L_{org}, e.g., DOM), and competing cations such as Ca^{2+} and excluded ligands (L_{ex}). Uptake of metal by organisms may be physiologi-

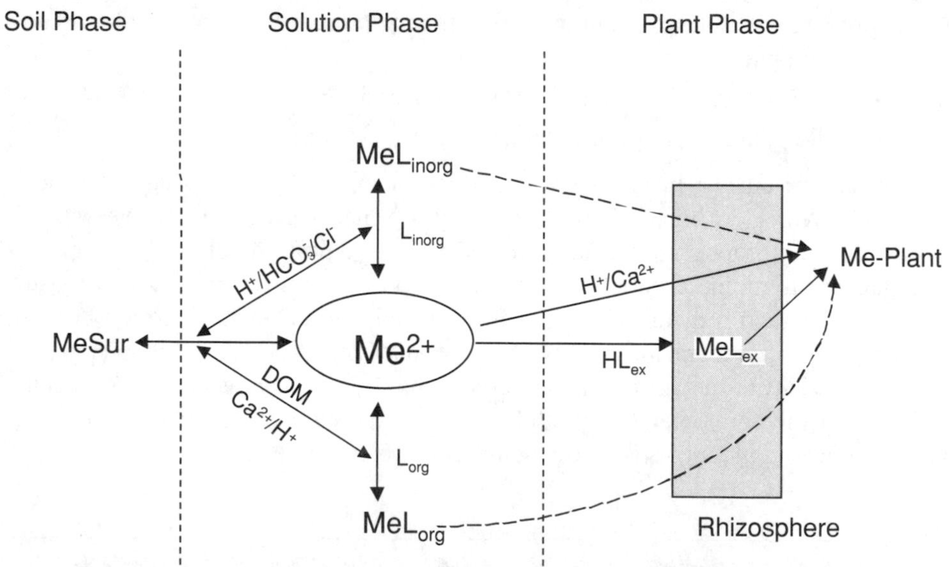

Figure 1-1 Environmental factors determining trace metal exposure to plants in soils

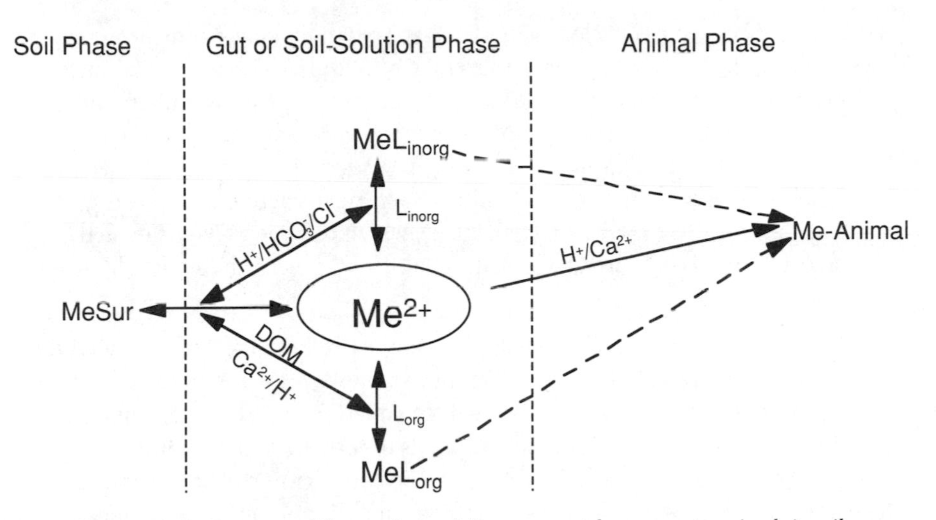

Figure 1-2 Environmental factors determining trace metal exposure to animals in soils

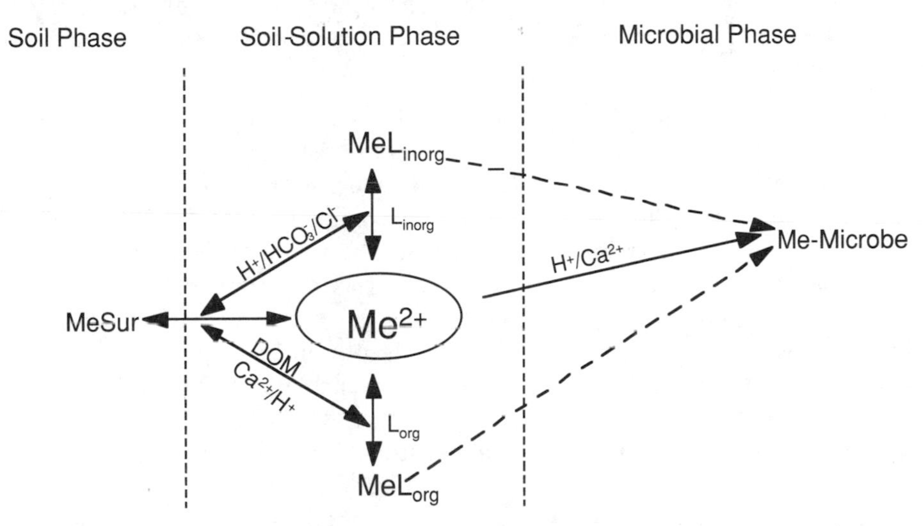

Figure 1-3 Environmental factors determining trace metal exposure to microbes in soils

cally moderated by pH and competing cations such as Ca^{2+}. Some inorganic and organic metal complexes may be directly taken up, while other ligands may compete with the organism for the metal.

Models in which the interaction of chemical species with organisms is used to predict trace metal uptake and toxicity were first introduced in aquatic toxicology by Pagenkopf (1983) and Morel (1983). Most commonly, this approach has been called the "Free Ion Activity Model" (FIAM), which recently has been reviewed by

Campbell (1995). Further development of the approach by Playle and coworkers (Playle et al. 1992, 1993a, 1993b; Janes and Playle 1995; Hollis et al. 1997; MacRae et al. 1999) has led to the understanding of the influences of protons, Ca ions, and organic matter complexation on the uptake of trace metals by fish gills and on the toxicity of the metals. This approach has been extended to a more comprehensive model capable of predicting toxicity to aquatic organisms across a variety of environmental conditions (Di Toro et al. 2001), which has been recommended (Hamelink et al. 1994) and proposed (Allen et al. 1999) for regulatory implementation.

In a few instances, the FIAM has been applied to the prediction of metal availability to higher plants. Parker and Pedler (1997a) have reviewed the potential limitations to the approach in soils. However, at present, the relationships between chemical forms of metals and the bioavailability of metals to terrestrial plants and to soil-dwelling organisms are not sufficiently understood to permit the prediction of uptake or toxicity of metals. We have used bioavailability in the same manner as Newman and Jagoe (1994) and Parametrix (1995) to indicate the degree to which the metal is taken up by the organism. This uptake may or may not affect the organism, but this is not considered in our usage. Because the metal concentrations of microbes growing in soil cannot be measured, we have used an effects-based endpoint, usually a toxic one.

Clearly, the determinants of metal bioavailability must be understood if one is to predict the effect of a metal. Soil screening levels and risk assessments should incorporate the essential factors that dictate the actual availability of metals. Such criteria could then validly be used to establish maximum tolerable levels of metal that can be accommodated in soil and as remediation standards. In the absence of this understanding of availability, regulatory agencies most commonly have used conservative default values. Present soil screening levels and standards usually are based on total concentrations of metals. Consequently, in considerations of sustainability, all inputs are considered to be potentially available, and only low-input amounts are considered to be acceptable. Assessment of soil quality frequently has driven concentrations to values less than background. Probabilistic relationships (Kooijman 1987; van Straalen and Denneman 1989; Aldenberg and Slob 1993) often are used to set criteria that are deemed protective of biological populations. Such statistical methods are important means to establish adequate levels of protection. However, they have validity only when the factors controlling bioavailability have been appropriately taken into consideration. Failing this, for any species, the studies that have been conducted in soils in which metals are the most highly bioavailable will always drive the assessments to low values. Because the input data are not adjusted to reflect bioavailability in the soil of concern, present criteria are usually conservative. However, it should also be noted that there would be a small proportion of sites in which the effects of the metal will be greater than predicted. In these instances, failure to incorporate bioavailability into criteria will result in underprotection.

It is the purpose of this review to summarize available literature regarding the determinants of partitioning of metals and the effects of partitioning on the availability of metals to microorganisms, plants, and soil-dwelling invertebrates. Based on our assessment, we have provided guidance as to means by which availability can be brought into soil screening levels. We indicate the approaches that should be taken to further this understanding to provide more fully developed criteria that adequately predict the effects of metals across a wide variety of soil types and metal concentrations. This assessment of the state of the science also provides important guidance for investigators to make valuable improvements to our understanding of metal bioavailability in terrestrial systems.

CHAPTER 2

Speciation of Metals in Soils

Sébastien Sauvé
University of Montréal

Metals are present in a number of forms in soils. Some forms or species are highly soluble, while others are so inert that their presence hardly influences the amount of the metal that is present in the soil solution phase. The distribution of these forms or species and their ability to partition to the soil solution are not constant; they vary with the time that the metal has been present in the soil. Aging of metals in soils tends to immobilize them and render them less available than freshly added metals. Consequently, in experiments in which metals have been added, the availability of metal is greater than in those in which the metal has aged for some period. Therefore, the availability of metals in these systems is often far greater than the same concentration of metal that has been aged. Likewise, the availability of metals freshly added as highly soluble metal salts exceeds that of metals added with a complexing matrix with high sorption capacity, such as sewage sludge (Logan and Chaney 1983).

The chemistry of soil, including that of trace elements, is discussed in standard texts on soil chemistry (Sposito 1989; McBride 1994; Sparks 1995). The mineralogy of soils will not be discussed in this book. Brümmer (1986), Evans (1989), McBride (1989), and Tomson et al. (2001) have published reviews of trace metal partitioning. Trace metals can be incorporated in the parent soil mineral matrix. This portion of the metal is generally poorly exchangeable and is released only by digestion with strong acids (e.g., aqua regia). Although the total content of trace metals in soils is frequently determined in both research and regulatory programs, it is clear that these values for trace metal content will not be useful in predicting the potential risk of the metals that have been determined. It is recommended that these procedures should not be used in programs where prediction of risk is the objective. For purposes of risk assessment, it is necessary to evaluate the concentration of metals in the soil that constitutes an exchangeable, potentially bioavailable pool.

Metals may also be present as secondary precipitates and as adsorbed phases. Secondary precipitation occurs when the solubility product of a metal compound is exceeded in the soil solution or at the soil surface. The treatment of soils that have been contaminated with Pb by the addition of phosphate to form insoluble Pb pyromorphites is an example of how metal precipitates can be used to immobilize Pb (Ruby et al. 1994). Mercuric sulfide has been reported (Barnett et al. 1995, 1997)

Bioavailability of Metals in Terrestrial Ecosystems: Importance of Partitioning for Bioavailability to Invertebrates, Microbes, and Plants.
Herbert E. Allen, editor. © 2002 Society of Environmental Toxicology and Chemistry (SETAC). ISBN 1-880611-46-5

in soils that have received Hg input. Metal hydroxide precipitates may form on the surface of soils and minerals at a lower saturation ratio than would be required in solution because of the effect of heterogeneous nucleation (Stumm 1992). A number of researchers have reported that metals precipitate on mineral and oxide surfaces at pH values much lower than what would be required for precipitation in solution (Scheidegger et al. 1997, 1998; O'Day et al. 1998).

Metals can be removed from solution by partitioning to inorganic and organic phases of soil. Reactive sites for the sequestration of metals occur on phases such as organic matter; oxides of Al, Fe, and Mn; and silicate mineral edge, planar, and interlayer sites. Numerous studies of sorption have been conducted on both individual soil components and whole soils. The mechanism for the retention of the metal by the partitioning phase is frequently unknown. The term "sorption" is used when the retention mechanism is unknown (Sparks 1995).

Determinations of Chemical Availability

The purpose of the Free Ion Activity Model (FIAM) or various soil extraction methods is to link simple and relatively inexpensive chemical determinations with the potential availability, the chemical reactivity, and the biological availability of possibly toxic elements. Determining chemical availability becomes critical because it is a much simpler and faster means of quantifying availability and allows an evaluation of bioavailability without the necessity to resort to a complete suite of bioassays. Unfortunately, "bioavailability" is an evasive concept. One suggested definition, which we have adopted, is "the amount or concentration of a chemical that can be absorbed by an organism thereby creating the potential for toxicity or the necessary concentration for survival" (Parametrix 1995). This definition reflects the duality of biological uptake, which integrates both the availability to satisfy a physiological requirement and the possible excess uptake leading to toxicological effects.

Experimental data have demonstrated that the bioavailability of a metal is highly dependent on the effects of soil physicochemical conditions (as reviewed in this book). This emphasizes the need for site-specific soil quality criteria because a similar moderate level of contamination might prove toxic and present a significant risk under a certain set of conditions (e.g., acidic sandy soil), whereas the same level of contamination might prove harmless in a different setting (e.g., calcareous clay). From a total-based generic criteria perspective, this implies that certain "polluted" soils may not present a significant health or environmental risk, and their forceful remediation would be a very expensive and a not very resource-efficient expenditure. On the other hand, this also means that certain soils otherwise considered only marginally contaminated (again, on a generic criteria basis) may actually show some significant environmental or health risks and warrant drastic efforts to minimize any metal burden increases or may require site remediation.

Free metal, extraction, and fractionation

An important distinction among the plethora of soil metal evaluation procedures is whether they are trying to determine the metal bound to the solid phase or that which is in the solution. For our purpose, the term "speciation" will be restricted to the determination of the chemical species in the soil solution, mainly a distinction among free metal species, inorganic ion pairs, and dissolved fulvic and/or humic complexes. Conversely, "fractionation" will be related to the compartmentation of the metal among the various solid phases present in the matrix (see Figure 2-1).

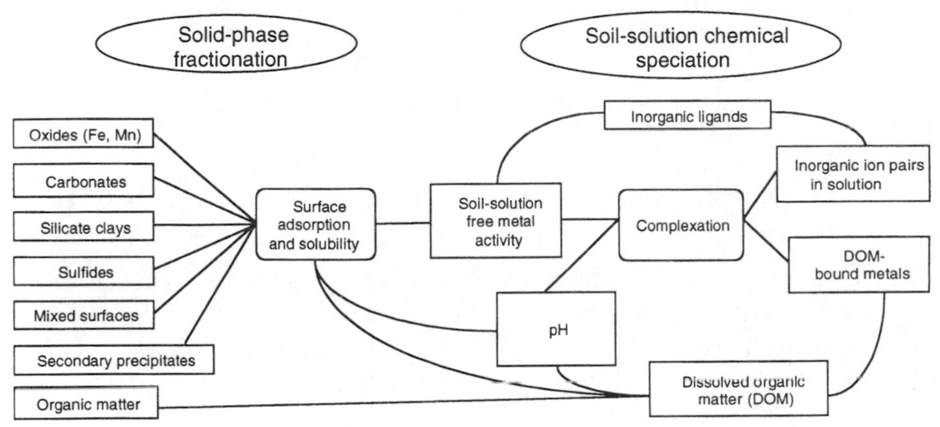

Figure 2-1 Representation of metal pools in solid-phase fractionation and components and reactions in soil-solution chemical speciation

Metal fractionation, the partitioning of the total soil metal associated with various solid phases, is usually estimated using specific operational extraction procedures or with the aid of physical methods such as spectroscopy (e.g., extended x-ray absorption fine structure [EXAFS] or x-ray diffraction).

Extractions, on the other hand, may represent an ill-defined mixture of metal compounds and complexes. Extractions and analysis methods range from field screening colorimetric determinations to sophisticated quantifications, as required to attain specific aims or objectives. Conceptually, certain soil extraction protocols aim at mimicking the conditions as they occur in the field. In those cases, the extractions become a means to determine the solid–solution partitioning of the metal of interest. This "partitioning" is often presented using a partition coefficient K_d, defined as

$$K_d = \frac{\text{Total Soil Metal}}{\text{Dissolved Metal}} \quad \text{(Equation 2-1)},$$

where total soil metal is in mg/kg and dissolved metal is in mg/L, hence K_ds are usually reported in L/kg. Although "total soil metal" usually refers to the adsorbed metal pool, in most situations this distinction is insignificant for trace metals.

This seemingly simple relationship is actually controlled by numerous complex processes. It is dependent on many physicochemical characteristics of the soil's solid and solution phases. Although this parameter, along with the total concentration of the metal in the soil, is frequently used to estimate potential risk, its limitations are evident. The metal in the soil can reside in both mineral phases and precipitates in addition to the more readily exchangeable adsorbed phase. If a generic or average value is used for K_d, the estimated value of metal in solution or risk will increase with increases in the amount of metal contained in the mineral phase as shown in the following equation:

$$[\text{Dissolved Metal}] = \frac{[\text{Total Soil Metal}]}{K_d}$$

$$= \frac{[\text{Mineral Phase Metal}] + [\text{Precipitated Metal}] + [\text{Adsorbed Metal}]}{K_d} \quad \text{(Equation 2-2).}$$

However, because metals in the mineral phase are relatively unavailable, this procedure often produces overestimates.

Therefore, it is recommended that total metal concentration in the soil not be used to determine partition coefficients that are to be used for risk assessment purposes. Many reviews recognize the limited applicability and relationships between metal bioavailability that are based on various extraction schemes (Beckett et al. 1983; Gunn et al. 1988; Beckett 1989).

Free Ion Activity Model

Metal bioavailability in aquatic systems is linked to the free metal activity in solution (Hart 1981; Allen 1993; Campbell 1995; Allen and Hansen 1996; Renner 1997). The same concepts can be applied to soils and show promising, but not yet conclusive, results (Bingham et al. 1983; Laurie et al. 1985; Minnich and McBride 1986; Minnich et al. 1987; Laurie and Manthey 1994; Knight and McGrath 1995; Parker et al. 1995, 1998; Sauvé et al. 1996, 1999; Sauvé, Dumestre et al. 1998; Dumestre et al. 1999; McGrath et al. 1999). But in opposition to aquatic studies, the properties of the solid phase and the kinetics of the reactions at the solid–liquid interface dominate soil solution chemistry. The application of the FIAM to soil systems is rendered more difficult by the complexity and the predominant influence of the processes of adsorption–desorption, precipitation–dissolution, uptake–release, and others that buffer and obscure the exchanges between the solid and the solution phases (Harter 1991b; Sheppard et al. 1992; Luoma 1995; Jin et al. 1996; Parker and Pedler 1997a; Peijnenburg et al. 1997). Furthermore, those various processes are very heterogeneous in nature: chemical (inorganic or organic) and biological. They also occur among the gaseous, liquid, or solid phases and hence are difficult to measure and quantify.

So, although most guidelines are based on total metals, mobility actually depends on dissolved metal concentrations and their chemical reactivity (complexation, adsorption, mineral equilibrium). Also, biological availability (whether through passive or active uptake mechanisms) occurs primarily through the soil solution. Most of these chemical and biological processes are mediated through free metal activities in the soil solution.

The fraction of the total soil metal dissolved in the solution in pristine or contaminated soils represents only about 0.001% to 0.01% for Cu and Pb, 0.05% to 15% for Cd, 0.001% to 5% for Zn, and 0.001% to 20% for Ni (Alloway et al. 1984; Gerritse and van Driel 1984; Sheppard and Thibault 1990; Bierman et al. 1995; Lorenz et al. 1997; Sauvé, Hendershot, Allen 2000). Even worse, the fraction of the total soil metal actually present as free metal ions in solution is between 10^{-8} and 10^{-10} for Cu and Pb and rises to between 0.002% and 13% for Cd (Sauvé 1999; Sauvé, Norvell et al. 2000). Therefore, variations in soil parameters among samples can lead to large differences in the concentration of soluble metal and even larger changes in the free metal ion activity.

Determination of metals in soils

Metal content

Determining total metal content of soil to measure potential availability is certainly the most conservative option. But even such a straightforward analysis is dependent on the analytical methodology. It is not the aim of this book to review the various methods or to do an exhaustive comparison of the techniques used for total metal digestions. Conceptually, "total metal analysis" refers to the complete dissolution of the solid matrix and quantification of its chemical constituents. Unfortunately, such analyses require a lot of analytical work for sample preparation and for the various acid and thermal treatment steps, and the preparation procedure often leads to analyte loss or sample contamination and yet may not achieve complete metal recovery (van Grieken et al. 1980; Binstock et al. 1991; Smith and Arsenault 1996). Alternatively, an approach aiming at the determination of total recoverable metals may be used; this simplifies the analytical work and improves reproducibility.

It is important to note that total soil metal estimates do not truly reflect the totality of the metals present; such discrepancy is especially pertinent in low background concentrations (see Chen and Ma 1998). The results of method comparison also show somewhat inconsistent results. Comparing microwave-assisted HNO_3 digestion to hotplate reflux digestion (using an HNO_3–HCl–H_2O_2 reagent) showed metal-specific biases of up to ±100% (Wei et al. 1997). In another study, as the sample metal concentration increased, the digestive ability of a procedure employing an aqua regia–HF mixture decreased relative to an HNO_3–$HClO_4$–HF digestion (Elliott and Shields 1988). Because the degree of contamination is often unknown for environmental samples, the harshest method is most reliable for assessing residual and true total metals in polluted soils.

On the other hand, from an ecotoxicological risk assessment perspective, irrespective of methodological details, the metal fraction, which is not dissolved in concentrated boiling acid (independently of the procedure), has a very low bioavailability, and the potential toxic contribution of such materials is very low. One problem that arises from the use of total metal in regulatory programs is that because risk may not be proportional to total metal, resources may be expended on sites of lesser importance. Nevertheless, it is important to make the appropriate considerations from a regulatory and risk communication point of view; that is, it is necessary to choose and recognize a specific reference method.

"Normal" soil metal levels are difficult to establish; they are clearly dependent on site history and even on global geochemical cycles. It is usually recognized that pristine soils not derived from unusual parent materials will show metal levels no higher than ~20 mg Cu/kg, ~1 mg Cd/kg, ~50 mg Ni/kg, ~25 mg Pb/kg, and ~50 mg Zn/kg (Canadian Council of Ministers of the Environment [CCME] 1991; Förstner 1995). Those levels are only representatives of somewhat arbitrary norms; a higher level locally may be perfectly "normal" and may not represent any real threat. Similarly, soils at those levels may not be protective of the environment under all possible site conditions. It is clear that potential environmental or human health risks need to be addressed by more discriminate soil metal determination techniques that are based on bioavailability considerations.

Adsorption–desorption

To evaluate the total soil metal that is in a bioavailable form, it is necessary to resort to soil extraction methods. Those usually use a desorption protocol of some sort. Even though it is often not identified as such, an extracting reagent will be used to desorb metals from the soil's solid phase. Such metal desorption reactions are rate dependent and will be affected by the kinetics of many chemical processes (desorption, dissolution, dissociation, complexation, competitive exchange, etc.). So, even if they are used as instantaneous measurements, most (if not all) extraction methods actually integrate an indication of the soil capacity to supply a certain element (metal release rate).

Soil solution

Predicting the properties of soil solutions is difficult, and even the "soil solution" itself has a range of definitions (Wenzel and Blum 1995; Fotovat and Naidu 1998). From a purist's point of view, the soil solution corresponds to what is obtained from in situ collection of tension cup lysimeters or other field-collected soil solutions. Unfortunately, even such samples are not impervious to methodological bias (ion sorption to the sampling interface device [Hendershot and Courchesne 1991; Wenzel et al. 1997], physicochemical and biological transformation of the sample in situ, after sampling, during storage, CO_2 degassing, etc.). Besides, the field installations required for such analyses are very resource intensive and impractical from most bioavailability testing perspectives. Nevertheless, such field research is needed to validate other soil extraction protocols aimed at mimicking "real" soil solutions.

The next protocol closest to in situ collection is probably the use of water displacement and centrifugation to obtain representative soil solutions (Ross and Bartlett 1990). Unfortunately, this is laboratory intensive and impractical for soils with high clay content (Lorenz, Hamon, McGrath 1994). The concentrations obtained in solution also depend on the moisture content of the collected soils and the concomitant effects on solution chemistry. Also, immiscible displacement is used but becomes even more complicated in terms of laboratory manipulations (Kittrick 1983; Elkhatib et al. 1986).

Another practical approach to obtaining soil solution is a simple batch-type extraction. In this case, the soil solution is assumed to be represented by the solution obtained from the shaking of a dilute salt soil extraction with a small solid–solution ratio (preferably 1:2 or less). Although this procedure is operationally defined, it is suggested as a surrogate of the soil solution because it occurs in the field while allowing for the collection of reasonable sample volumes with a minimum of laboratory manipulations. It also has the advantage of maintaining a known and fixed concentration of counterions (which is required, or appreciated, for many free metal speciation techniques). Often, the centrifuged solution is filtered through cellulosic or polycarbonate membranes (Florence 1982; Shaw et al. 1984; Jopony and Young 1994). It is usually recognized that this separates particulate from dissolved components (although this is also a somewhat arbitrary definition because smaller-sized colloids will not be removed).

Although it would be nice to have a simple, single definition of soil solution and how it should be obtained, the achievement of such a definition is not felt to be possible. Researchers have different needs and different biases that must be respected.

Water and neutral salts

Among the various extraction techniques, water is the simplest alternative. As such, centrifugation of field moist solutions or extraction by displacement will yield a soil solution whose characteristics are closer to those in the field than is a soil solution derived by water extraction. But it is important to realize that distilled deionized water is not a natural product. From this point of view, weak salt extractions (also called "exchangeable") may prove to be better surrogates of soil solutions than pure water (because soil water may contain between 0.002 to 0.0001 M Ca [Fruchter et al. 1990; Hendershot and Courchesne 1991]). The absence of counterions in pure water yields metal-specific desorption variations and drastically affects the solubility of organic matter as well as the complex metals associated with dissolved organic matter (DOM).

Many neutral salt extractions have been suggested and compared (Häni and Gupta 1982, 1985a, 1985b; Gupta 1984; Salt and Kloke 1985; Haddad and Evans 1993; Ure et al. 1993). The concentrations vary from a low of 0.001 M to a high of 1 M salt solutions of $CaCl_2$, $Ca(NO_3)_2$, KNO_3, $NaNO_3$, NH_4NO_3, etc. Although it would be important to standardize a soil-solution extraction protocol, it also is important to realize that specific research needs vary. The ideal extraction reagents will differ according the specific needs of each study. For example, surrogate extraction solutions for arid-zone alkaline and sodic soils would demand a high ionic strength and ought to be dominantly Na-based. Solutions for moderately acidic forest soils would need a low ionic strength, and the predominant cation ought to be Ca or Al. Using NO_3 anions may seem appropriate to reduce Cl complexation and electrochemical interference but would not be possible in conjunction with nitrification (or other N-transformation) bioassays.

It is vital to emphasize the importance of the cation and anion salt component, and the "best" choice is probably the one most representative of the soil solution in the studied ecosystem. As such, water or neutral salt extractions are closely linked to determinations of partitioning coefficients (K_d), which are discussed in more detail in "Partitioning" (p 17).

Specific extractants and sequential fractionation

Single extractant procedures are widely used to estimate metal availability to plants, for instance, to establish the quantity of fertilizer that should be applied. Among the extractants that have been used are diethylenetriaminepentaacetate (DTPA), ethylenediaminetetraacetic acid (EDTA), acetic acid, and the mineral acids HNO_3 or HCl (Adriano 1986). However, when applied to soils with a broad spectrum of metal concentrations or soil characteristics, they fail to adequately predict metal bioavailability.

The original concept of soil (or sediment) fractionation procedures was appealing: Various reagents were specifically chosen to dissolve a certain solid phase and liberate the associated metals, starting with the most accessible metals and sequentially using harsher reagents to remove more recalcitrant metal fractions (Tessier et al. 1979; Lake et al. 1984; Tack and Verloo 1995). Ideally, this would allow the identification of the solid phase to which the metals are associated. Theoretically, different chemicals can remove metals associated with different fractions of the soil. For instance, $CaCl_2$ or $MgCl_2$ presumably extracts metals found in the soluble and exchangeable fractions of the soil (McLaren and Crawford 1973a; Janzen 1993); DTPA and EDTA are believed to extract the exchangeable and organically bound trace metals (Trierweiller and Lindsay 1969; Norvell 1984; O'Connor 1988) and also to dissolve metal "precipitates" (Schalscha et al. 1982). Na-acetate is used to specifically target the carbonate-bound metal fraction (Tessier et al. 1979). Hydroxylamine hydrochloride and citrate dithionite have been used to attempt to identify the metals associated with the "reducible" solid phase and to quantify metals bound to Fe, Al, and/or Mn oxides (Chao 1972; Shuman 1985). An exhaustive review of sequential fractionation methods is not warranted because the reagents used cannot truly identify the suggested soil pools. At best, sequential extractions identify operationally defined soil fractions, and none can be consistently related to soil metal bioavailability. For example, the extractions are not specific; they extract more or less than the fraction they are supposed to extract, and as the extraction sequence proceeds, some of the metals redistribute themselves among the remaining solid-phase fractions (Jouanneau et al. 1983; Tipping et al. 1985; Rapin et al. 1986; Kheboian and Bauer 1987; Martin et al. 1987; Bauer and Kheboian 1988; Tessier and Campbell 1988; Nirel and Morel 1990; Kim and Fergusson 1991; Xiu et al. 1991; Bermond 1992, 1993; Xiao-Quan and Bin 1993; Qiang et al. 1994; Whalley and Grant 1994; Flores-Vélez et al. 1996; Ma and Uren 1998).

In any case, there is no clear evidence indicating which solid-phase fractions correlate best with uptake by soil organisms. Also, the difficulty in choosing an extractant is compounded by its lack of universality in predicting biological uptake of metals. For instance, numerous studies have demonstrated that the effectiveness of an extractant to correlate with plant uptake is metal dependent and is specific to plant species, plant part, and growth stage (e.g., Haq et al. 1980; Roca and Pomares 1991; Davies 1992; Taylor et al. 1992; Haddad and Evans 1993).

Furthermore, attempts to use "reagent-specific" extractions as estimates of bioavailable metals yield variable and inconsistent results (e.g., Gunn et al. 1988; Beckett 1989; Cook and Hendershot 1996; Qian et al. 1996). To cite a World Health Organization working group on the risks of land application of sewage sludge: "In spite of numerous attempts, no extractant solution has been found that can be used to predict the availability of even one element to a wide variety of plants..." (Dean and Suess 1985). Because sequential fractionation procedures and various extractions fail to identify chemically specific soil pools and are not good estimates of bioavailability, their usefulness is very limited and other approaches are needed.

Uptake by organisms other than plants may involve ingestion of soil particles. In this case, the solution in which the desorption of metals from the particles occurs may be quite different from that of soil pore water. For human exposure, Ruby et al. (1996) have developed a physiologically based extraction system in which the soil is first exposed to conditions simulating those in the stomach, followed by those encountered in the intestinal tract. One can envision that similar procedures could be useful in defining the availability and metal uptake by soil-dwelling organisms.

Chelating agents and dilute acids

Chelating extractions using DTPA and EDTA were initially developed to identify trace element crop deficiency in slightly acidic and alkaline soils (Lindsay and Norvell 1969a, 1978; Norvell 1984). To quote Norvell (1984): "The limitations of the DTPA micronutrient soil test under strongly acid and metal-enriched soil conditions suggest that a different procedure might be more suitable." So, although the original procedure was successful in its original design, the wider applications to assessment of contaminated soils may not be appropriate (O'Connor 1988).

Chelating agents and dilute acid extractions represent an intermediate potential between "total" digestions and water extractions. This approach presumes that the metal fractions that are available for dissolution by the extractant are also potentially bioavailable in the long- to mid-term. Although the chelate extraction, along with a plethora of variants (Sauerbeck and Rietz 1982; Lake et al. 1984; Beckett 1989; Soon and Abboud 1993), has had some surprising success for soil testing from an agronomical perspective, its application to soil contamination evaluation has had mixed results. Again, as discussed for sequential fractionations, no single extraction has given consistent results for multiple metals and/or multiple species (Cook 1997; McLaughlin et al. 2000).

Spectral techniques

Extended x-ray absorption fine structure is an alternative method for the identification of the solid-phase associations of metals. EXAFS and other spectral techniques allow the use of reference materials to fingerprint the structural environment of a trace metal in complex matrices like soils (Teo 1986; Fulghum et al. 1988; Essington and Mattigod 1991; Flores-Vélez et al. 1996; Manceau et al. 1996; O'Day et al. 1998). The required synchrotron facilities are not widely available, and although the approach is useful to fractionate metals in contaminated mineral materials, it may be unable to characterize the matrix of a number of metals in soils at realistic contamination levels (i.e., below 1000 mg/kg). Again, and most importantly, the solid-phase associations identified this way have not yet been linked to bioavailability. Nevertheless, the application of those techniques with the specific aims of distinguishing between metal adsorption and co-precipitation could prove useful in determining the relative release rates of metals, in terms of bioavailability through both soil solution and gastrointestinal transit (Ruby et al. 1992, 1999; Hamel et al. 1998). Similarly, multiple spectroscopic approaches can also be applied more directly on organic matter (Senesi 1992).

Physical fractionation

Physical fractionations can be used to localize the soil fraction where the metals can be found (Yeoman et al. 1989; Ugolini et al. 1996). Usually, metals are located within the finer soil fractions or the lighter (organic-rich) phases (Tessier et al. 1982; Ducaroir et al. 1990; Essington and Mattigod 1990; Christensen 1992; Ducaroir and Lamy 1995; Flores-Vélez et al. 1996). Although the physical separation is relatively simple, the information gained is also limited. Nevertheless, it is essential to have a very good idea of the localization of the soil metals before a soil remediation project is implemented, whether for size reduction or for the choice of the most appropriate methods (Peters and Shem 1995). Magnetic and density gradient separations also are used to identify the components of street dusts and to provide estimates of relative bioavailability (Biggins and Harrison 1980).

Partitioning

The partitioning of metals is partly dependent on how it is determined. Some of the principal factors to consider include the solid–liquid ratio, the time of extraction, the shaking or centrifugation methods, the phase separation and filtration, and the water or extraction solution properties. It is further necessary to account for the physicochemical characteristics of the soil to model the site-to-site variability observed for various K_d determinations.

Extraction methods

Solid-to-solution ratio

The solid–liquid partitioning of metals in contaminated soils has been studied and reported under various experimental designs. One of the simplest models remains the K_d coefficient (see Equation 2-1). But even such a simple model requires certain assumptions for its application. For example, the solid–liquid partitioning is affected significantly by the solid–liquid mass ratio; some experiments in aquatic systems use ratios as low as 1:100 or even less. The relative adsorption evidently is reduced over the partitioning of a field soil with 20% moisture (i.e., a solid–liquid ratio around 5:1). Some experimental studies evaluate the relative importance of solid–liquid ratios on metal sorption (Skyllberg 1995; You et al. 1999; Yin et al. 2002). The largest solid–liquid ratio practically feasible is recommended as a model system of field-occurring solid–liquid partitioning. In practical terms, this has meant the use of solid–liquid ratios between 1:0.8 and 1:2; anything lower is a serious technical challenge, and higher ratios are not representative of soil solutions (excessive dilution of DOM). It is important to realize that if a fixed quantity of a chemical (whether an exchangeable metal pool or DOM) is available for dissolution, a lower solid–liquid ratio dilutes that chemical, and the results obtained from that experiment are not truly representative (You et al. 1999). This "dilution" is a complex process, dependent on many physicochemical parameters. The effect of time of extraction was explored for certain metals (Brümmer et al. 1988; Sauvé et al. 1995), but those results are metal specific and cannot be extrapolated to other metals without experimental verification.

Sample preparation

Sample preparation is also very important because air-drying (and even more so, oven-drying) will change the form and the speciation of the soil metals and will drastically alter the properties of the soil organic matter (SOM) (Bartlett and James 1980; Wenzel and Blum 1995). The hydrophobic character of organic matter sometimes may slow the return of the wetted soils to a physicochemical state similar to the pre-drying conditions. Drying also will change the chemical and possibly the mineralogical properties of oxides (Fe, Al, and Mn) and is an unacceptable procedure for chemically reduced (anaerobic) samples because this will result in oxidation of sulfide.

Estimation of K_d

Some K_d data for metals were compiled in Sheppard and Evenden (1988b), Sheppard and Thibault (1990), McBride et al. (1997), and Sauvé, Hendershot, Allen (2000). Much of the research shows that although the K_d approach, which uses a single ratio to distribute the metal pool between the solution and the solid phases, is simple to implement and model, its application is limited mostly to the system that was used for its calibration. Modeling metal partitioning under varying chemical conditions requires consideration of the properties of the solution and the solid phases (Römkens and Salomons 1993).

Various methods are available to estimate default K_d values that are based on partitioning ratios between soil and plants (Baes and Sharp 1983; Sheppard and Evenden 1988a, 1992, 1996; Sheppard and Sheppard 1989, 1991; Sheppard et al. 1989; Sheppard 1991) or based on extrapolations of relationships of chemical properties (Sheppard and Sheppard 1989; Sheppard and Thibault 1990).

Regression models

It is difficult to segregate the effects of various physicochemical soil properties, but various regression models have been suggested in the literature (Abd-Elfattah and Wada 1981; Barrow et al. 1981; Sanders 1982; Brümmer et al. 1986; Reddy and Dunn 1986; Anderson and Christensen 1988; Buchter et al. 1989; Neite 1989; Duquette and Hendershot 1990; Allen et al. 1994; Gooddy et al. 1995; Janssen et al. 1996; Lee et al. 1996; Janssen, Peijnenburg et al. 1997; Janssen, Posthuma et al. 1997; McBride et al. 1997; Radovanovic and Koelmans 1998; Temminghoff et al. 1998; Elzinga et al. 1999; You et al. 1999; Sauvé, Hendershot, Allen 2000). It seems that the main factors needed are pH and adsorptive characteristics such as organic matter, oxide content, or cation exchange capacity (CEC). The K_d value is also dependent on the actual level of the contamination, on the levels of competing ions in solution, and on DOM.

Solution pH

Soil solution pH is definitely the predominant factor influencing metal solubility, but it is difficult to discuss its influence independently from other physicochemical characteristics because so many processes are interrelated. For example, although Cu desorption is increased at lower pH, the acidity reduces the concentrations of DOM and Cu-hydroxy complexes (Brümmer et al. 1986; Temminghoff et al. 1998). Furthermore, solution cation concentrations are dependent on pH, will affect conformational changes of DOM, and may provoke its coagulation (Sposito and Holtzclaw 1977; Römkens and Dolfing 1998).

Cations

It is also important to note that a dilute salt (such as 0.01 M $CaCl_2$) extracts less of some metals than does pure divalent water (Gerritse and van Driel 1984; Pardo 1997; Fotovat and Naidu 1998). Because salt cations (e.g., Ca^{2+}) promote the flocculation of DOM, they will provoke the precipitating of the associated metals with a strong affinity for the DOM (e.g., Cu and Pb) (Saar and Weber 1979, 1980b; Gamble et al. 1984; Escrig and Morell 1998; Fotovat and Naidu 1998; Römkens and Dolfing 1998).

On the other hand, cations in solution may competitively reduce the adsorption of divalent trace metals (Gupta and Harrison 1981; Christensen 1984, 1987; Lorenz, Hamon, McGrath et al. 1994; Echeverría et al. 1998). So, it is difficult to intuitively predict the relative effects of variations in solution cation levels because different processes are competing among themselves. We need to consider cation competition for exchange sites on the solid phase (with different relative effects on organic versus mineral sites), cation competition for complexation to dissolved ligands, flocculation effects on organic matter dissolution, cation influences on dissolution reactions, as well as the strong mediating effect on the kinetics of various chemical reactions for some of these processes.

Inorganic anions

Complex and ion-pair formation of metals with inorganic anions is easier to model than is complexation with organic ligands. This arises partly from a better understanding of the inorganic chemical equilibrium reactions and the complexity and heterogeneity of humic- and fulvic-metal complexation. (Chemical equilibrium model calculations are discussed in more detail in "Chemical Equilibrium Models," p 27.)

Although the ion-pair formation of metal-inorganic anions is relatively straightforward, the potential environmental implications are very complex and difficult to foresee. For example, the application of phosphate fertilizers to reduce the solubility and bioavailability of Pb in contaminated orchard soils results in desorption of arsenate (Davenport and Peryea 1991) and simply replaces one problem with another (and a more mobile toxicant in this case). It is also unclear to what extent organic and inorganic ligands compete together for adsorption on the soil anion exchange sites, and this also yields complex interactions that are due to the association of metals on some of these ligands.

Organic matter

Organic matter generates complex interactions because of the duality of the effects arising from DOM versus organic matter in the solid phase. SOM contributes to the CEC and will increase metal adsorption. This reduces solution metal concentrations, with a higher preference for those metals having a greater affinity for organic ligands. On the other hand, that same affinity for organic ligands will increase metal complexation in the soil solution, thereby reducing adsorption and increasing dissolved metal concentrations. It is interesting to note that the relative affinity of certain metals may vary between dissolved and solid organic binding sites. For example, Saar and Weber (1980b) have shown that Pb has a high affinity for flocculating DOM, thereby shifting the relative affinity of organic matter for Pb, Cu, and Cd. To further complicate things, divalent cations binding to DOM may promote its coagulation and generate some complex conformational changes in the properties of organic matter (Sposito and Holtzclaw 1977).

Although increasing solution pH will decrease metal mineral solubility, it may increase dissolved metal concentrations because of concomitant formation of metal complexes with DOM as well as carbonate and hydroxy species. Also, DOM may coat reactive mineral adsorption sites, inactivating them and therefore indirectly inhibiting metal adsorption (Harter and Naidu 1995). In contrast, organic ligands might sorb on the mineral phase, thereby generating new sorbing surfaces and increasing metal retention (Petruzzelli et al. 1992). It is clear that many interactions are possible and that experimental data are necessary to validate any model (Temminghoff et al. 1998).

Also, partitioning is frequently studied through the use of adsorption isotherms, particularly to develop partitioning constants. The maximum adsorption capacity is frequently determined this way, and the strength of binding can be determined from the slope. Yin, Allen, Huang, and Sparks (1997) showed that for soils with higher amounts of organic matter, the adsorption isotherms were not the conventional L-shape but were S-shaped. This is a result of the initial amounts of the added metal (Hg) reacting with the DOM, and only after that capacity was exceeded was there appreciable sorption to the solid. Again, this demonstrates the importance of organic matter to the partitioning and of relating laboratory procedures to actual field conditions.

The importance of organic matter in the partitioning of other metals has also been studied by Yin et al. (2000). They investigated desorption of the Cu, Ni, and Zn contained in the same 15 soils that had first been used by Lee et al. (1996). The logarithms of the distribution coefficients for dissolved Ni and Zn (but not for dissolved Cu) were dependent on pH. All distribution coefficients were highly dependent on the soil–solution ratio. The partitioning of Cu followed that of the soil organic C. Although the logarithm of the distribution coefficient for soluble Cu was not correlated to the pH, the logarithm of the distribution coefficient for the free Cu ion was highly dependent on the pH ($R^2 = 0.80$). When the logarithm of the distribution coefficient for the Cu ion was normalized to the content of organic matter in the soil, the value of R^2 increased to 0.96.

Compilations of experimental data

A considerable amount of data is reported in the literature for solid-to-liquid partitioning of metals (McLaren and Crawford 1973b; Frost and Griffin 1977; MacLean and Dekker 1978; Tiller et al. 1979; Elsokkary 1980; Gerritse and van Driel 1984; Zhan 1986; Sheppard et al. 1987; Sieghardt 1987; Xian 1987; Campbell and Beckett 1988; Vuori et al. 1989; Joshi et al. 1991; Abuzid and Obukhov 1992; Kubota et al. 1992; Levy et al. 1992; Petruzzelli et al. 1992; Trueby et al. 1992; Jeng and Singh 1993; Pierzinsky and Schwab 1993; Allen et al. 1994; Mench et al. 1994;

Ramos et al. 1994; Atanassova 1995; Winistörfer 1995; Camobreco et al. 1996; Lee et al. 1996; Sauvé et al. 1996; Krishnamurti et al. 1997; Janssen, Posthuma et al. 1997; Li 1997; Ma and Uren 1997; Sauvé, McBride, Hendershot 1997; Brun et al. 1998; de Groot et al. 1998; Esnaola and Millán 1998; Holm et al. 1998; Kalbitz and Wennrich 1998; Knight et al. 1998; Lebourg et al. 1998; Richards et al. 1998; Yamada et al. 1998; McGrath et al. 1999; Sauvé 1999; Ge et al. 2000; Sauvé, Hendershot, Allen 2000; Tambasco et al. 2000). A thorough review of fluxes of Cd, Cr, Cu, Ni, Pb, and Zn in forests is also available (Bergkvist et al. 1989).

A compilation of metal solubility data (Sauvé, Hendershot, Allen 2000) reported in the literature (references cited above) for extracted soil solutions (using many varied extraction protocols) gave average values (±SD) for "uncontaminated soils" as follows:

4 ± 9 μg Cd/L,
39 ± 43 μg Cu/L,
108 ± 360 μg Ni/L,
14 ± 23 μg Pb/L, and
520 ± 970 μg Zn/L.

Although the definition of "contaminated" is somewhat arbitrary, the wide range of concentrations of a metal that are encountered should be noted.

Using that same compilation, it can be seen that there is a significant influence of pH, total metal burden, and sometimes SOM on solid-to-liquid partitioning. The gathered data are illustrated in Figure 2-2, and the resulting regressions are given in Equations 2-3 to 2-7 from data compiled by Sauvé, Hendershot, and Allen (2000).

$$\log(K_d - \text{Cd}) = 0.48\ \text{pH} + 0.82 \cdot \log(\text{SOM}) - 0.65 \quad \text{(Equation 2-3)},$$
$R^2 = 0.613, n = 751, p < 0.001.$

$$\log(K_d - \text{Cu}) = 0.21\ \text{pH} + 0.51 \cdot \log(\text{SOM}) + 1.75 \quad \text{(Equation 2-4)},$$
$R^2 = 0.419, n = 353, p < 0.001.$

$$\log(K_d - \text{Ni}) = 1.02\ \text{pH} + 0.80 \cdot \log(\text{SOM}) - 4.16 \quad \text{(Equation 2-5)},$$
$R^2 = 0.758, n = 69, p < 0.001.$

$$\log(K_d - \text{Pb}) = 0.37\ \text{pH} - 0.44 \cdot \log(\text{Total Pb}) + 1.19 \quad \text{(Equation 2-6)},$$
$R^2 = 0.562, n = 204, p < 0.001.$

$$\log(K_d - \text{Zn}) = 0.60\ \text{pH} - 0.21 \cdot \log(\text{Total Zn}) - 1.34 \quad \text{(Equation 2-7)},$$
$R^2 = 0.573, n = 298, p < 0.001.$

It can be seen that the predictive strength of the various regressions varies from 42% to 76% (for corresponding R^2) on log-log scales. Much of the variability in this data set must be attributed to differences in the experimental protocols among the various studies. The methodological variability blurs the true variability of the soil relationships. This variability is compounded by variations in methods for both the dissolved (soil solution extraction) and particulate metals pools (total metal diges-

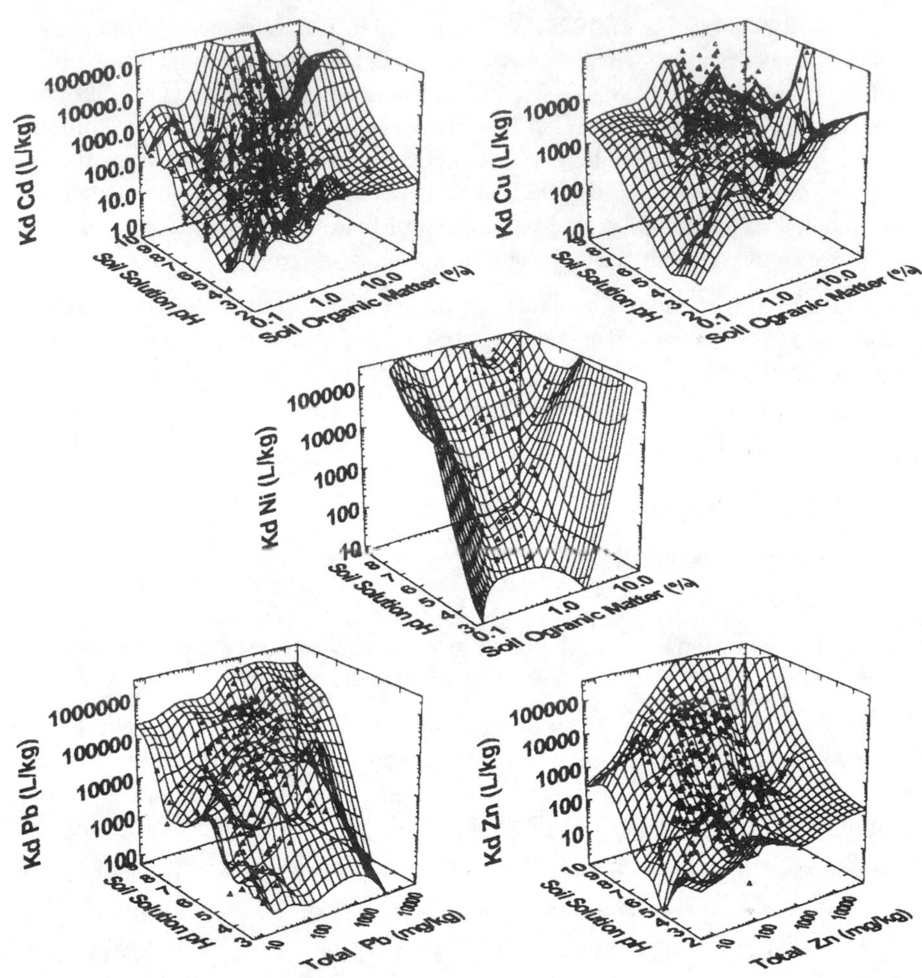

Figure 2-2 Compiled K_d values as a function of soil solution pH and either log SOM (as %C for Cd, Cu, and Ni) or log soil total metal (in mg/kg for Pb and Zn) (Drawn from data compiled by Sauvé, Hendershot, Allen 2000.)

tion). Nevertheless, it is interesting to note that the pH coefficients vary from 0.21 for Cu to 1.02 for Ni. This reflects the slope of the pH response, that is, for a given change in pH, the solubility of Cu is much less affected than is that of Ni or Zn (with Cd and Pb in an intermediate position). A lack of pH sensitivity of soil-solution dissolved Cu concentrations has also been observed earlier (Sauvé, McBride, Norvell, Hendershot 1997; Salam and Helmke 1998; Yin et al. 2002).

It is also notable that the effect of variations of the total metal burden is negligible for Cd (coefficient is 0.06), that Cu and Ni have similar coefficients (~0.3), that Pb is higher, and that Zn is lower. The overall effects of those relationships are partly described in Figure 2-2. Data from solutions obtained from field-collected soils and from pH-adjusted batch equilibration studies are included. It often has been observed that freshly added metal salts are much more soluble than are field-aged contaminants. Still, the in situ–collected samples are more representative and valuable, combining as much data as available to increase the sample number and improve the predictive strength of the resulting regressions. Given enough incubation, salt-spiked samples may approach a pseudo-equilibrium that is close to that of natural soils. Experimental work is still needed to establish when a laboratory- or field-contaminated soil reaches steady state with respect to metal solubility and chemical speciation. A significant portion of the variability within that data set arises from the differences among the experimental protocols, and the data are weighted according to the respective sample size, with a relatively high influence impact of one study within the Cd regression (Lee et al. 1996). The regressions are reported in a K_d format for easy inclusion within many readily available environmental fate models, the aim being to predict dissolved metal concentrations. Different regression approaches are discussed in more detail elsewhere (Sauvé, Hendershot, Allen 2000).

Analytical Determinations of Soil-Solution Free Metal Activities

Various techniques can be used to measure the free metal activity in the soil solutions, each with its own advantages and limitations (see Florence and Batley 1980; Astruc et al. 1981; Florence 1982, 1986; Apte and Natley 1995 for reviews).

Electrochemical methods

The most direct method for determining free metal species is possibly through ion-selective electrodes (ISEs) — in which case few, if any, assumptions are required for the interpretation of the results (Saar and Weber 1980a; Florence 1986). There is a generous literature on their application (Blaedel and Dinwiddie 1974; Buffle et al. 1977; Bresnahan et al. 1978; Sekerka and Lechner 1978; Gamble et al. 1980; Wagemann 1980; Bhat et al. 1981; Maeda et al. 1981; Umezawa et al. 1981; Gulens et al. 1984; Hart and Jones 1984; Kerven et al. 1984; Stella et al. 1984; Verney et al. 1984; Cabaniss and Shuman 1986; McGrath et al. 1986; Mench et al. 1987; Vlasov and Bychkov 1987; Zhukov et al. 1988; Morrison and Florence 1989; Holm 1990; Camusso et al. 1991; Town and Powell 1993; Pungor 1996). Furthermore, the manipulations are simple, similar to pH measurements. In the case of Cu, the electrode is not prone to excessive ionic interference (Cavallaro and McBride 1980; Florence 1986; Gulens 1987; Sauvé et al. 1995) except possibly high chloride concentrations (Westall et al. 1979; Belli and Zirino 1993). The Cu ISE has been used to determine soil solutions activities down to 10^{-12} M Cu^{2+} (Sauvé et al. 1995; Logan et al. 1997). It has also been used in pure synthetic-buffered solutions to measure activities as low as 10^{-19} M Cu^{2+} (Hansen et al. 1972; Avdeef et al. 1983). The ISE is the method of choice for Cu.

In the case of Pb, the ISE can be used for synthetic solutions of known chemical composition down to free Pb^{2+} activities of 10^{-10} to 10^{-12} M (Hansen and Rùzicka 1974; Kivalo et al. 1976; Sauvé, McBride, Hendershot 1998a) and also in peat and humic acid (HA) extracts (Logan et al. 1997). But for soil solutions, the Pb ISE is prone to excessive interference from Fe and/or organic matter.

Differential pulse-anodic stripping voltammetry (DP-ASV) is another very sensitive electrochemical method and can be used to determine the concentrations of "labile" metal (Florence and Batley 1980; Florence 1982, 1986; Iyer et al. 1989; Morrison and Florence 1989; Opydo 1989; Waller and Pickering 1990; Pinheiro et al. 1994; Deaver and Rodgers 1996). The free Me^{2+} activity in the extracted soil solutions can be estimated by assuming that this DP-ASV labile fraction represents easily dissociable inorganic ion pairs and excludes organically complexed metals (Figura and McDuffie 1979; Florence 1986; Shuman 1988; Pinheiro et al. 1994). The measured ASV-labile metal is partitioned into the expected ion pairs using the formation constants for the inorganic reactions of interest in soil solutions (Sauvé, McBride, Hendershot 1997, 1998b; Sauvé 1999; Sauvé, Norvell et al. 2000) or in iron oxide suspensions (Gonçalves et al. 1985; Palmqvist et al. 1997).

Some studies have shown that the presence of inorganic or organic colloids does not interfere with the ASV measurements (Gonçalves et al. 1985, 1987; Aualiitia and Pickering 1986). But it is reported that the adsorption of organic matter on the Hg-drop electrode may hinder metal diffusion and thus diminish the current readings (Florence 1986). As such, "it is often difficult to determine if a metal wave is diminished because of physical interference to diffusion, by formation of inert organo complex or by a combination of the two processes" (Florence 1986). Florence also emphasizes the importance of calibrating by using peak-heights for ionic metal, rather than attempting to quantify the results by standard additions. Smart and Stewart (1985) have used membrane-covered electrodes to test the relative voltammetric interference of various organic surfactants. Capelo et al. (1995) have used a Nafion-coated Hg-film electrode, and Kubiak and Wang (1989) have tried to eliminate the problem by using fumed silica amendments to "purify" the solutions. It is important to realize that DOM may interfere in the voltammetric determination of "labile" metals. Such measurements therefore are dependent on the use of calibration solutions with properties similar to the samples and on the assumption that the main effect of DOM is to complex metals without directly influencing the Hg interface (Pinheiro et al. 1994).

Nonelectrochemical methods

Competitive chelation — the addition of a strong chelate to soil solutions — has also been used for the speciation of free divalent metal ions in soil solutions (Lindsay and Norvell 1969b; Norvell and Lindsay 1969a, 1969b; Santillan-Medrano and Jurinak 1975; Lindsay 1979; Fujii et al. 1983; Hendrickson and Corey 1983; Kalbasi et al. 1995). In this case, the authors assume that an added solid phase maintains a constant free metal activity of a second reference metal and that the chelate addition does not by itself modify the surface solution equilibria (and the speciation). Its use is limited to nonacidic and preferably alkaline soils, to minimize the mass action interference from competing metals that would be solubilized at low pH. This approach also was used for the determination of free Cd^{2+} activity in soil solutions (Fujii et al. 1983; Workman and Lindsay 1990; El-Falaky et al. 1991; Ma and Lindsay 1995), Cu^{2+} (Norvell and Lindsay 1969a, 1969b; Fujii et al. 1983; Chaudhri 1993), Zn^{2+} (Lindsay and Norvell 1969b; Fujii et al. 1983; Sachdev et al. 1992; Ma and Lindsay 1993), Pb^{2+} (Kalbasi et al. 1995), and Ni^{2+} (Fujii et al. 1983; Ma and Lindsay 1995).

Exchange resins are also used to measure the free metal activity in soil (Camerlynck and Kiekens 1982; Dietze and König 1988; Lee and Zheng 1994; Apte and Natley 1995; Liang and Schnoenau 1995; Fotovat and Naidu 1997). This method was applied to the determination of free Cd^{2+} and Zn^{2+} in soil solutions (Holm et al. 1995, 1998; Lorenz et al. 1997; Fortin and Campbell 1998) and for Pb^{2+}, Cu^{2+}, Cd^{2+} (Lee and Zheng 1994), and Ni^{2+} (Dunemann et al. 1991). Ion exchange resins can also be used to determine the relative lability of metals (Beveridge et al. 1989; Slavek et al. 1990; Waller and Pickering 1991, 1992; Lee and Zheng 1993; Procopio et al. 1997; Esnaola and Millán 1998), but this does not necessarily provide an estimate of the free metal activity; it sometimes is used only to give an approximation of relative lability. Furthermore, using exchange resins in this way necessitates a laboratory-intensive characterization of the adsorption properties of the resin of interest under the proposed chemical conditions.

Donnan dialysis membranes also can be used to determine the speciation of soil solutions (Dietze and König 1988; Berggren 1989; Apte and Natley 1995). Minnich and McBride (1987) have used this technique to determine the free Cu^{2+} activity in Cu salts and sewage sludge–amended soils. Comparison with the Cu ISE showed that the ion activities measured by the 2 techniques were well correlated, even though the electrode was more precise and the Donnan dialysis membrane technique underestimated activity by as much as 1 log unit relative to the Cu ISE (Minnich and McBride 1987). Salam and Helmke (1998) have also used Donnan membranes for the solution speciation of free Cd^{2+} and Cu^{2+} in contaminated soils. The Cd speciation results compared very well to those obtained using ASV (Sauvé, Norvell et al. 2000), but the results obtained for the membrane speciation of Cu^{2+} overestimated activity relative to ISE measurements (Sauvé 1999). Berggren (1989) also applied the combined use of dialysis and ion exchange resin procedures to the speciation of Cd, Cu, and Pb, as well as for many different elements (Dietze and König 1988).

Comparison of speciation techniques

Scant studies are available for comparison of free metal speciation techniques in natural soil solutions. Amacher (1984) discusses some of the limitations of ISEs and ion chelate equilibria, and Sparks (1984) and Sposito (1984) discuss some of the theoretical and historical aspects of ion activity determinations. It is important to realize that it is difficult to use synthetic ligands of known affinities to directly validate a speciation technique for DOM complexation. DOM has much slower reaction kinetics than do simple organic molecules (Ma et al. 1999); the best model for calculating its resulting equilibrium metal association is still debated because it is prone to coagulation and steric effects, and it is very difficult to quantify the extent of multidentate binding.

Using synthetic solutions for which the experimental results could be compared with chemical equilibrium calculations, ISEs and ASV gave good results (McGrath et al. 1986; Shuman 1988; Morrison and Florence 1989). On the other hand, when Cu^{2+}-fulvic acid systems were used, it was found that the ISE and the aluminum hydroxide resin gave results concordant with an algal assay, but ASV yielded complexation capacity (a measure of the concentration of excess complexing agents; Florence 1989) more than 1 order of magnitude higher than the other techniques. This was attributed to an incapacity of ASV to discriminate partially labile Cu complexes with a large complexation capacity. A similar excess of complexation capacity for Cu was found for ASV relative to the ISE for soil solution extracts (Sauvé 1999). A comparison of chelating resin and ASV has shown that the resin method overestimates metal lability relative to ASV (Figura and McDuffie 1979). An interesting comparison of speciation techniques applied to Cu complexation in groundwater showed that, relative to the ISE, fluorescence quenching underestimated and cathodic stripping voltammetry overestimated the degree of Cu^{2+} binding (Holm 1990; Holm and Curtiss 1990). Also, Fortin and Campbell (1998) used an ion exchange technique to show a positive interference for free Cd^{2+} and Zn^{2+} determinations in the presence of synthetic amino-acid ligands (e.g., lysine and cysteine).

For soil solutions, Dietze and König (1988) and Berggren (1989) have compared ion exchange resin procedures and dialysis speciation, and both have both found that equilibrium dialysis was appropriate and that ion exchange resins gave measures of relative lability that were not equivalent to free metal. For soil-derived fulvic acid, ASV was found to agree well with ISE measurement, provided appropriate calibrations were undertaken (Bhat et al. 1981). For aquatic systems, ISE and fluorescence quenching agreed well under certain conditions, but the ISE was more reliable. The "polydisperse mixture character" of natural organic acids needs to be incorporated into the evaluation of data (Gamble et al. 1980; Cabaniss and Shuman 1986).

Semimechanistic predictive regressions (see "Free Metal, pH, and Soil Total Metal," p 33) were used to illustrate the differences between the prediction results relative to the reported measurements from the various soil-solution speciation studies (Figure 2-3). In almost all cases, the various nonelectrochemical techniques yielded free metal activities up to 3 orders of magnitude higher than what is predicted from the model. Very often, the comparison showed a strong pH-dependent bias. The only good fit of the predictive regression models with nonelectrochemical speciation was observed with Donnan membrane speciation of Cd^{2+} (Salam and Helmke 1998).

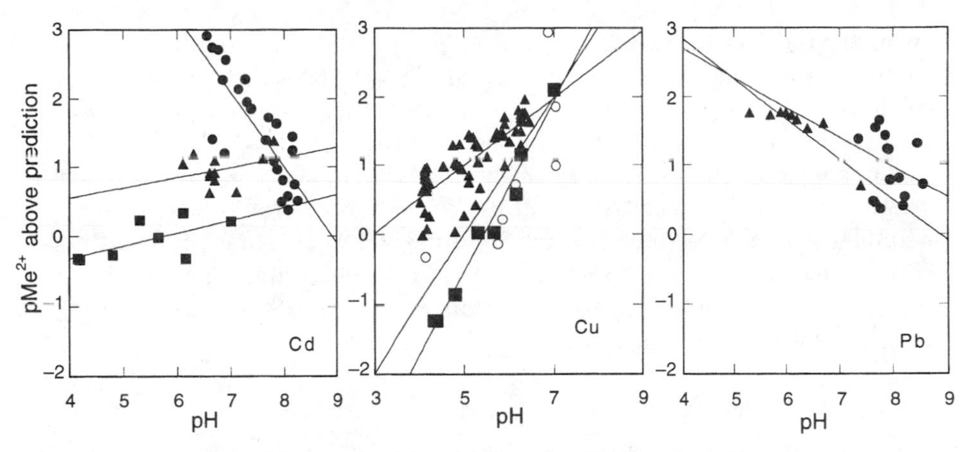

Figure 2-3 Comparison of free metal activity pMe^{2+} predicted by Equation 2-12 to reported measurements of soil-solution free metal activity speciation using exchange resins (▲; Lexmond 1980; Holm et al. 1995, 1998; Lebourg et al. 1998), competitive chelation (●; Workman and Lindsay 1990; El-Falaky et al. 1991; Kalbasi et al. 1995; Ma and Lindsay 1995), DP-ASV Cu^{2+} (○; Sauvé 1999), and Donnan membrane (■; Minnich and McBride 1987; Salam and Helmke 1998). Lines represent linear regressions through the data from each study.

The lack of fit between speciation methods is not a true comparison (predictive regressions are used to compare estimates against measurements). Furthermore, some of the experimental differences may actually arise from variations in the methodological extraction and equilibration protocol, irrespective of the speciation technique. The predictive regressions were developed from a data set using a relatively short (20 minutes) 1:2 soil:0.01 M $CaCl_2$ or KNO_3 extraction protocol. Other studies often used longer equilibration times. Part of the difference may arise from the sample extraction protocol, and an unquantified part of the discrepancy may not be attributed to the speciation techniques (nevertheless, the pH-dependent biases can hardly be attributed to the extraction protocol).

Experimental results are needed to characterize the effects of sampling protocol variations. It is also critical to properly compare the performance of various free-ion activity speciation techniques under identical conditions.

Chemical Equilibrium Models

Estimating free metal activities

Chemical equilibrium models are not speciation techniques per se. Models can be used to calculate the chemical speciation of a solution of known composition, but this is a computation of the speciation, not a measurement. In some cases, care is taken to consider and estimate complexation by DOM, but too often this is simply neglected. The calculated speciation is dependent upon the appropriateness of the chemical modeling equations and constants used, as well as the accuracy of the chemical input data. In the case of hydrolysis products and inorganic ion pairs in solution, the formation constants are known fairly well (Table 2-1), and models should be quite accurate (McGrath et al. 1986; Parker et al. 1995; Turner 1995). Some results for the inorganic chemical speciation of Cd, Cu, Pb, and Zn in soil solutions are illustrated in Figure 2-4. Figure 2-4 is built from the predictive model equations for dissolved metal (similar to Equations 2-3 to 2-7) and those for free metal activity (see "Free Metal, pH, and Soil Total Metal," p 33) and the formation constants compiled in Table 2-1. In the case of Cd, the main inorganic ion pairs are $CdCl^+$ and $CdSO_4^0$. For Cu, the main inorganic species are $CuSO_4^0$ and some hydroxide complexes at higher pH.

Table 2-1 Stability constants for inorganic speciation of free metal activity in soil solution[a]

Equation	Cd log *K*	Cu log *K*	Ni log *K*	Pb log *K*	Zn log *K*
$Me^{2+} + H_2O \Leftrightarrow MeOH^+ + H^+$	−10.08	−7.70	−9.86	−7.71	−8.96
$Me^{2+} + 2H_2O \Leftrightarrow Me(OH)_2^0 + 2H^+$	−20.35	−13.78	−19.00	−17.12	−16.90
$Me^{2+} + 3H_2O \Leftrightarrow Me(OH)_3^- + 3H^+$	−33.30	−26.75	−30.00	−28.06	−28.40
$Me^{2+} + HCO_3^- \Leftrightarrow MeHCO_3^+$	2.10	2.70	2.20	2.90	2.10
$Me^{2+} + CO_3^{2-} \Leftrightarrow MeCO_3^0$	4.12	6.75	6.87	6.27	4.75
$Me^{2+} + 2CO_3^{2-} \Leftrightarrow Me(CO_3)_2^{2-}$	6.40	10.69	10.11	9.49	9.63
$Me^{2+} + NO_3^- \Leftrightarrow Me(NO)_3^+$	0.31	0.50	0.40	1.17	0.40
$Me^{2+} + 2NO_3^- \Leftrightarrow Me(NO_3)_2^0$	0.00	−0.40	−0.60	1.40	−0.30
$Me^{2+} + Cl^- \Leftrightarrow MeCl^+$	1.97	0.40	0.40	1.58	0.49
$Me^{2+} + 2Cl^- \Leftrightarrow MeCl_2^0$	2.59	−0.12	0.96	1.82	0.62
$Me^{2+} + 3Cl^- \Leftrightarrow MeCl_3^-$	2.40	−1.57	—	1.71	0.51
$Me^{2+} + SO_4^{2-} \Leftrightarrow Me(SO)_4^0$	2.30	2.36	2.29	2.62	2.33

[a] Sources: Lindsay 1979; Smith and Martell 1989; Schecher and McAvoy 1991; Evans et al. 1995; Lumsdon et al. 1995; Carroll et al. 1998.

In the case of Pb, the main inorganic species are $PbSO_4^0$ and $PbCl^+$ with very significant contributions of $Pb(OH)_2^0$ and $PbCO_3^0$ at circumneutral pH and above. For Zn, only the hydroxide species becomes important at pH above 7.5.

For the divalent metals in Figure 2-1, the increased total dissolved metal concentrations at the higher pH values reflect complexation by DOM. Furthermore, in soils, it is critical to distinguish between the solution and the dominant solid phase (Figure 2-1). Speciation in soil solutions also requires consideration of ion exchange processes and chemical sorption onto a solid phase that has been subjected to widely

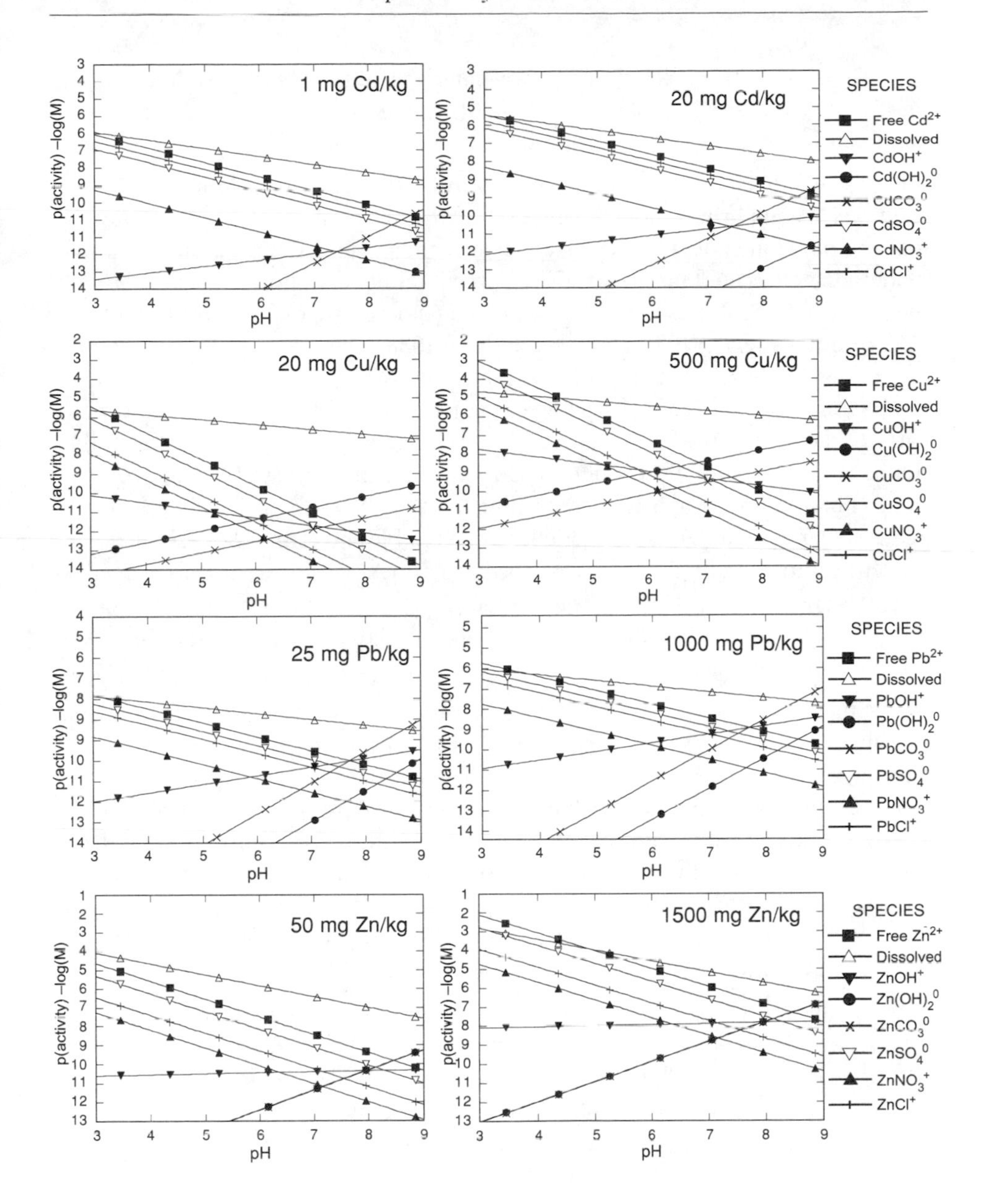

Figure 2-4 Dissolved metal species in soils contaminated with Cd, Cu, Pb, or Zn. Solubility of Cd in 2 soils contaminated with 1 and 20 mg Cd/kg. Free Cd^{2+} is estimated using Eq. 2-13. Cu in 2 soils contaminated with 20 and 500 mg Cu/kg. Free Cu^{2+} is estimated using Eq. 2-14. Pb in 2 soils contaminated with 25 and 1000 mg Pb/kg. Free Pb^{2+} is estimated using Eq. 2-15. Zn in 2 soils contaminated with 50 and 1500 mg Zn/kg. Free Zn^{2+} is estimated using Eq. 2-16. Inorganic ion pairs are calculated using the constants in Table 2-1. Dissolved Cd, Cu, Pb, and Zn are estimated using the pH-dependent regressions from a K_d compilation (Sauvé, Hendershot, Allen 2000). The p(activity) and –log(M) are for the activity of M, the indicated metal species.

variable pedogenetic processes and is very heterogeneous on both the micro- and macroscales. It is also important to consider the potential impact of inorganic anions such as chloride, nitrate, and sulfate (Doner 1978).

Chemical equilibrium models can be used to predict trace metal solubility of solutions in equilibrium with pure minerals with good accuracy. But soils are not made of pure minerals, and they may not be at equilibrium. Chemical equilibrium models do not account for the kinetics of precipitation–dissolution of solids or for the slow reaction rates of certain trace-metal ligand exchanges (Allen et al. 1982; Plankey and Patterson 1987; Hering and Morel 1988, 1989, 1990a, 1990b; Harter 1991a; Hering 1995; Ma et al. 1999). Certainly, the solubility of pure mineral phases can be compiled easily and incorporated into a chemical equilibrium model. Various authors have shown the free metal activity produced by various minerals at equilibrium over a wide range of pH (Lindsay 1979; Stumm and Morgan 1996; Sauvé 1999). But in soils, the processes controlling trace metal solubility in a heterogeneous solid phase (like soils) are complex and variable and cannot be described simply by mineral solubility. There is much experimental evidence that free metal activities are maintained below the levels required for mineral equilibria, and the free metal/pH slope is usually less than the slope of 2 that is predicted for stoichiometry of divalent metals.

Modeling metal complexation by DOM is more complicated than modeling mineral equilibria. Various conceptual binding models have been suggested for solutions, most of which have been designed for aquatic systems. Some models assume discrete multisite or multidentate binding, while others use a continuous binding site concept (electrostatic, normal distribution, or affinity spectrum). Most model comparisons agree that while the continuous distribution models are theoretically more sound, a stochastic approach to determine the optimum pH-dependent multisite binding is more appropriate. Turner (1995) says this about nonelectrostatic independent binding sites models:

> On a purely pragmatic basis this model must be judged a great success. It is simple to apply, it gives excellent fits to experimental data, and it is fully compatible with existing speciation modeling packages. However, totally empirical models such as these cannot be used to extrapolate beyond the range of conditions under which the experimental data were obtained.

Other chemists have reached similar conclusions when they compare discrete and continuous binding models (Buffle 1980; Cabaniss et al. 1984; Dzombak et al. 1986; Fish et al. 1986; Martell et al. 1988).

More sophisticated computational models also account for polyelectrolyte properties and functional group heterogeneity (Ephraim et al. 1986; Ephraim and Marinsky 1986a, 1986b; Kramer et al. 1990); further corrections also can be included to account for electrostatic attraction–repulsion and nonspecific ion binding that results from counterion accumulation (Tipping and Hurley 1992; Tipping 1994).

It must be remembered that these models are designed for aquatic systems, although once properly calibrated, they are very useful for extracted soil solutions. Still, they were not designed a priori to accommodate the very large solid-phase component of soils, the accompanying buffer capacity, and the complexity of a solid phase in dynamic relation (not equilibrium) with a transient soil solution, which has ever-changing concentrations of mineral and organic particulates.

Competitive adsorption model

Soil-solution free metal activity can be linked to soil properties by some very simple soil properties such as pH, organic matter content, and total metal content. Many regressions have been suggested to predict metal solubility, but few of them actually predict free metal activities. Many predict dissolved metal concentrations or "labile" fractions, either ignoring or presuming negligible metal complexation (Jopony and Young 1994; Janssen et al. 1996; Janssen, Peijnenburg et al. 1997; de Groot et al. 1998; Elzinga et al. 1999).

McBride et al. (1997) and Sauvé (1999) have used a simple semimechanistic regression model to predict the free metal speciation of soil solutions with various origins (also see Jopony and Young 1994). This simple, mass action–driven complexation model assumes that metal (Me) complexation by a deprotonated adsorptive surface (Sur) with y dissociable protons bonded to the surface can be described by

$$Me^{2+} + SurH_y \Leftrightarrow MeSur + y\,H^+ \quad \text{(Equation 2-8).}$$

This can be transformed into a conditional formation or stability constant K, written as

$$K = \frac{[MeSur](H^+)^y}{(Me^{2+})\,[SurH_y]} \quad \text{(Equation 2-9),}$$

where square brackets represent concentrations of MeSur (surface-bound metal) and $SurH_y$ (protonated surface sites), and (H^+) and (Me^{2+}) denote the ion activities of H^+ and Me^{2+} in solution. Transforming Equation 2-7 to the logarithmic form yields

$$pMe^{2+} = -\log Me = \log K - \log\left(\frac{[MeSur]}{[SurH_y]}\right) + y \cdot pH \quad \text{(Equation 2-10),}$$

where pMe^{2+} is the molar activity of the Me^{2+} free metal ion. Furthermore, assuming that the fraction of the metal adsorption capacity actually bonded with trace metals is negligible ($MeSur < SurH_y$), then $SurH_y$ can be related to the total metal-reactive adsorptive surfaces of the soil. Equation 2-10 can then be transformed by adding distinct coefficients and parameters for metal loading and reactive adsorptive surfaces. This yields a linear regression of the following form:

$$pMe^{2+} = a + b \cdot pH + c \cdot \log(\text{Total Metal}) + d \cdot \log(\text{Sur}) \qquad \text{(Equation 2-11)},$$

where a, b, c, and d are coefficients determined by statistical regressions and appropriate sets of data. Total metal is the soil metal content measured by acid digestion, and Sur is the capacity of the surface sites represented by organic C content or another measure of surface exchange sites.

Equation 2-11 may assume that trace metal adsorption is controlled by the organic fraction of the soil. Although this may seem reasonable given the high affinity of organic ligands for trace metals, alternatively it is possible that in certain soils and for some metals (e.g., Pb), the inorganic or mineral fraction has a significantly greater influence or even control over solubility, regardless of the organic matter content. In those situations, it may be reasonable to replace the organic matter term in Equation 2-11 with a measure of the clay or Fe and Al content or even a less specific measure of total adsorption capacity (such as CEC or specific surface area). Then, it is tempting to add and try multiple linear regressions using all the available measured soil parameters, but it is important to realize that most soil characteristics are autocorrelated to various degrees (Basta et al. 1993), and such regressions including all sorts of data would yield a best fit equation of limited usefulness. Furthermore, the relative importance of the soil solution pH is such that, for data sets of mineral soils of various origins, a statistically significant regression can be obtained without even introducing the quantities of soil adsorbents (i.e., clays, oxides, organic matter) as parameters (Sauvé, McBride, Norvell, Hendershot 1997; Sauvé 1999; Sauvé, Norvell et al. 2000). This is equivalent to assuming that various soils contain a similar average concentration of adsorptive surface sites or assuming that the actual effects of varying adsorptive site concentrations are negligible, compared with the variability controlled by pH and total metal levels. Alternatively, this also could be explained, assuming that the metal adsorptive properties of soils are dependent mostly on the organo-mineral coatings of the soils. Those coatings represent a small fraction of the soil mass, but they are very reactive; they are the most exposed surfaces and might be similar in soils of different origins. In effect, if the term $SurH_y$ from Equation 2-10 is set to a constant, Equation 2-11 is reduced to a simpler relationship because $SurH_y$ is itself pH sensitive, and its variability may already be factored into the pH variable to some degree:

$$pMe^{2+} = a + b \cdot pH + c \cdot \log(\text{Total Metal}) \qquad \text{(Equation 2-12)}.$$

Equation 2-10 has been successfully applied to the soil-to-solution partitioning of Cd^{2+}, Cu^{2+}, Pb^{2+}, and Zn^{2+}, whereas using Equation 2-11 (including the exchange capacity – SOM term) could be justified statistically only in the case of Cu^{2+} and Cd^{2+} speciation. Although this seems to imply a small importance of SOM, it is important to note that the soils in those data sets were not very rich in organic matter (SOM < 100 g C/kg soil), and the inclusion of soils high in organic matter changed the regressions (McBride et al. 1997). The much higher concentrations of metals (e.g., Cu, Zn) necessary to induce phytotoxicity in crops grown in organic soils compared to mineral soils suggest a critical importance of SOM in the control of solution free metal activity. Under controlled soil characteristics (e.g., pH, clay size fraction origin, particle size distribution) the importance of SOM in controlling metal activity and solubility is more obvious (Lee et al. 1996; McBride et al. 1997; Sauvé, McBride, Norvell, Hendershot 1997; Sauvé, McBride, Hendershot 1998b; Sauvé 1999; Yin et al. 2002) and has been demonstrated in many studies.

Free Metal, pH, and Soil Total Metal

The soil-solution free metal activity of the various metals can be predicted using pH and soil total-metal content, as given in Equation 2-10. Although the resulting coefficients are different for Cd, Cu, Pb, and Zn (see "Experimental data relating pH and soil total metal to free metal," p 33), the relationships are similar for the 4 metals: a clear log-linear increase in the free Me^{2+} activity in response to an increase in total metal content and a log-linear increase in divalent metal activity in response to increased soil solution acidity. Similar regressions also have been derived and used in other studies (Jopony and Young 1994; McBride et al. 1997). Furthermore, experimental measurements of soil-solution free metal activities have shown that the metal levels needed for metal activity control by mineral solubility equilibrium are rarely reached, and then only for very drastically contaminated soils, for example, >10,000 mg Cd/kg or >15,000 mg Cu/kg or >5,000 mg Pb/kg (Jopony and Young 1994; Sauvé, Martinez et al. 2000). This supports the conceptual model that free metal activity is controlled by complexation–adsorption reactions, as opposed to mineral solubility equilibria.

It is also notable that a single semiempirical model (Equation 2-12) is successful in predicting much of the variation in free metal speciation of Cd^{2+}, Cu^{2+}, Pb^{2+}, and Zn^{2+} in more than 100 field-collected soils of different origins and with various sources of metal contamination. Although only 62% to 85% of the variability is explained by the model, the predicted parameter is the most sensitive and critical parameter, the soil-solution free metal activity, instead of more easily predicted parameters such as the fraction of the total metal adsorbed (Wen et al. 1998).

It is possible that part of the success of the competitive adsorption model (Equations 2-11 and 2-12) is due to the use of total recoverable metal determined by concentrated HNO_3 digestions instead of total dissolution using HF digestion. The portion of the soil metal content that is not released by HNO_3 is very unreactive and, from an environmental perspective, may have a very low potential impact.

Experimental data relating pH and soil total metal to free metal

Speciation data for soil solutions were collected and compiled for soils contaminated by Cd, Cu, Pb, and Zn to various levels and with the contamination originating from different sources. In each case, field-collected soils were extracted for 20 minutes with a dilute salt solution (0.01 M KNO_3 for the speciation of Cd and Pb, and 0.01 M $CaCl_2$ for Cu) in 1:2 soil-to-solution mixtures. Potential metal losses during the filtration of the solutions were deemed negligible (Florence 1982; Jopony and Young 1994). Although this extraction method is not ideal (see "Extraction methods," p 17), it has allowed the processing of a large number of soils and the derivation of preliminary regressions.

The soils used for the derivation of the free Cd^{2+} regression come from a data set made of field-collected soils containing from 0.1 to 38 mg Cd/kg. The soil solution extracts were filtered with 0.22-µm cellulosic membranes and speciated using DP-ASV (Sauvé, Norvell et al. 2000).

$$pCd^{2+} = 5.14 + 0.61 \cdot pH - 0.79 \cdot \log(\text{Total Cd}) \quad \text{(Equation 2-13)},$$
$$R^2 = 0.696, n = 64, p < 0.001,$$

where Total Cd is in mg/kg.

The free Cu^{2+} regression is derived from a data set of field-collected soils speciated with an ISE (Sauvé, McBride, Norvell, Hendershot 1997; Sauvé 1999). The Cu concentrations go from 14 mg/kg to 16,000 mg/kg.

$$pCu^{2+} = 3.20 + 1.47 \cdot pH - 1.84 \cdot \log(\text{Total Cu}) \quad \text{(Equation 2-14)},$$
$$R^2 = 0.921, n = 94, p < 0.001,$$

where Total Cu is in mg/kg.

Experimental data for deriving a similar regression for Ni are not available.

The data for the free Pb^{2+} regressions come from a set of field-collected soils speciated by DP-ASV of extracts filtered through 0.22-µm cellulosic membranes (Sauvé, McBride, Hendershot 1997). The concentrations vary from 10 mg/kg to 15,000 mg/kg.

$$pPb^{2+} = 6.78 + 0.62 \cdot pH - 0.84 \cdot \log(\text{Total Pb}) \quad \text{(Equation 2-15)},$$
$$R^2 = 0.643, n = 84, p < 0.001,$$

where Total Pb is in mg/kg.

A preliminary predictive regression for free Zn^{2+} can be obtained by combining some ASV speciation results for urban contaminated soils (Tambasco et al. 2000) and for a pH-adjusted orchard soil from Ithaca, New York, USA (Sauvé, unpublished data) as well as some ion exchange resin speciation results for various European agricultural soils (Knight et al. 1998).

$$pZn^{2+} = 4.70 + 0.95 \cdot pH - 1.71 \cdot \log(\text{Total Zn}) \quad \text{(Equation 2-16)},$$
$$R^2 = 0.760, n = 30, p < 0.001,$$

where Total Zn is in mg/kg.

Soil organic matter's role in speciation and partitioning

Although pH is clearly the dominant factor controlling the distribution of metals between the soil and the solution phases, the effects of other soil and solution properties cannot be ignored in a more comprehensive treatment. Although many soil studies have focused on partitioning, insights into the soil solution speciation can be gained from those works. Lee et al. (1996) studied the adsorption of Cd onto a set of 15 soils that were adjusted over a wide pH range. The concentration of Cd added was much larger than the complexation capacity of the solution phase. The effect of pH was clearly the dominant factor controlling the partitioning. When the partition coefficients were normalized to the content of organic matter in the soil, the correlation significantly improved and soil phases other than the organic matter were unimportant in the binding of the Cd. They further suggested that the common practice of multiple linear correlation of the partition coefficient with pH and soil properties such as organic matter, metal oxide, and clay contents was invalid because the ability of the organic matter and the metal oxides to sorb metals is pH dependent.

Yin and coworkers studied the partitioning of Hg to the same set of soils used by Lee et al. (1996). The adsorption of low concentrations of Hg(II) as a function of pH did not follow the conventional adsorption edge pattern of low sorption of metal at a low pH rising to a plateau of nearly all metal sorbed at high pH values (Yin et al. 1996). They found that nearly all the Hg was adsorbed at pH values of 3 to 5. As the pH increased, less Hg was adsorbed. This decrease was due to the complexation of the Hg(II) with organic material dissolved from the soil. Because only 10^{-7} M Hg had been added, and because Hg forms strong complexes with organic matter, even small amounts of DOM had important influences on the partitioning process. The impact of organic material was more important than that of chloride ions on the partitioning, except at very low pH for soils with low organic matter content.

The rate of supply of metal from the solid phase to the solution phase is an important consideration in the understanding of metal availability. Yin, Allen, Huang, Sparks, and Sanders (1997) showed that both the adsorption and desorption rate coefficients were related inversely to the organic matter content of the soils studied. Furthermore, not all of the adsorbed Hg could be desorbed. The fraction of the Hg that was retained by the soil increased as the organic matter content of the soils increased.

Linking the Free Ion Activity Model to bioavailability

The application of the FIAM to soil systems is not as simple as its application to aquatic systems. Nevertheless, the derivation of soil quality criteria can be adapted to integrate some of the chemical aspects that have a determinant impact on metal bioavailability (de Haan et al. 1987; Allen 1993; van Straalen et al. 1994; Peijnenburg et al. 1997; Sauvé, Dumestre et al. 1998). There is some experimental evidence that, indeed, free metal is a valid predictor of bioavailability in metal-contaminated soils (Minnich and McBride 1986; Minnich et al. 1987; Laurie and Manthey 1994; Sauvé et al. 1996; Parker et al. 1998; Dumestre et al. 1999; McGrath et al. 1999). The challenge here is to be transparent about the limitations of this approach and also to present the science clearly, so that legislators and stakeholders may perceive the benefits and problems associated with the use of such a risk-based assessment.

Under certain circumstances, elemental supply may be controlled by factors other than the free metal activity in the solution. It must be kept in mind that many of the soil properties are autocorrelated to various degrees, and it is often difficult to segregate the actual contribution of distinct factors (Basta et al. 1993). Indeed, it is expected that the best predictor of metal availability will not always be identified as the free metal activity because of simultaneous correlations with other soil factors.

More specifically, there is ample evidence that chloride has a determinant impact on plant Cd uptake, that is, for a given free Cd^{2+} activity, more Cd will be taken up under higher-solution chloride concentrations (Smolders et al. 1998; McLaughlin, Maier et al. 1999). Similarly, Fe is often supplied as an EDTA complex to increase its availability (actually, to prevent its precipitation out of solution). In other circumstances, metal transport or diffusion to the site of uptake might control availability.

One of the weaknesses of predicting metal bioavailability from free metal activities in soil solution is the lack of consideration of the dynamic processes responsible for maintaining that free metal activity, as well as a somewhat simplified view of biological uptake (Parker and Pedler 1997a).

Plant elemental uptake is controlled by chemical availability in the soil solution as well as the soil's capacity to supply that element (Nye 1966; Nye and Marriott 1969; Barber 1984; Bell et al. 1991; Laurie et al. 1991b; Laurie and Manthey 1994). Most plant tissues will accumulate many times the amount of metal available in the soil solution at any given moment; effectively, the soil solution is "emptied" and replenished many times within even a single day (Bouldin 1989). So plant metal uptake in this case not only is dependent on the availability of the metal in the soil solution ("intensity") and the particularities of that plant's uptake mechanisms but also on the soil's capacity to supply that particular element ("capacity"). Some of these intensity–capacity concepts have been explored for plant nutrient uptake, but they are more difficult to identify in the case of toxic trace metals (Cd or Pb). The specifics of each biological species need to be linked to environmental availability. It is likely that the understanding of metal bioavailability (and impact) in soils will require the consideration of both the intensity of the toxic exposure (through the FIAM) and the soil's capacity to maintain this level of free metal activity in solution (desorption from soil surfaces, dissociation from complexed ligands, and diffusion-controlled element supply).

Conclusions

Soil extractions and sequential extractions do not provide universal estimates of bioavailability. Instead, alternative models are needed to incorporate the effects of solution free metal speciation upon bioavailability. Some simple semimechanistic models are available for predictions of soil-solution free metal activities of Cd^{2+}, Cu^{2+}, and Pb^{2+}, and some preliminary data are available for Zn^{2+}. The FIAM is proposed as an interesting alternative approach, although some refinements are necessary to improve the predictive ability to account for rate-dependent metal uptake. Furthermore, the biological uptake of trace elements is complex because different species have evolved specific requirements and methods to acquire, maintain, and control micronutrient uptake (such as Cu or Zn) and distinct mechanisms to restrict uptake of nonessential trace elements such as Cd and Pb.

Experimental comparisons of free metal speciation techniques are necessary for validation of extraction protocols with field-collected solutions, as well as for validation of various speciation techniques. Furthermore, there are too few experimental studies trying to link free metal speciation and various bioassays (many of the published experiments do not even report essential information such as soil pH or total recoverable metal contents). In much of the speciation or bioassay research, chemical and biological studies are done separately, and that research needs to be integrated and combined. It will then be possible to use the speciation results to explain and understand the observed biological responses and to justify the need to resort to the somewhat more resource-intensive chemical determinations of free metal activities.

CHAPTER 3

Bioavailability of Metals to Terrestrial Plants

Michael J. McLaughlin
CSIRO Australia

A key pathway for metal exposure to animal species (including humans) results from the uptake by plants of elements from soil in which terrestrial plants grow. Understanding metal uptake by terrestrial plants is therefore critical to assessing risks posed by soil metals to ecological and human agricultural food chains. Nonfood-chain risks of soil metals also may be modified by plant growth processes; for example, revegetation of bare soil areas with elevated metal concentrations can reduce direct exposure of metals to terrestrial animals.

Uptake and translocation of metals by plants is also important in relation to movement within soil profiles, for example, from subsurface to surface soil layers, which may thus markedly modify exposure pathways for terrestrial fauna. Historically, the study of metal uptake by plants has focused on micronutrient metals important in agricultural production, that is, Cu, Co, Mn, Mo, Ni, and Zn. Nonessential metals (Cd, Hg, and Pb) generally have received less attention, although Cd has been the subject of several investigations over the last 20 years because of its potential for food-chain bioaccumulation. More recently, metal hyperaccumulator species have been used to reexamine mechanisms of metal uptake by plants in light of the potential for phytoremediation of metal-contaminated soils.

The hazards to the environment from metal uptake by plants can be categorized as follows:

1) introduction of metals into the food chain,
2) loss of vegetation cover induced through phytotoxicity, and
3) cycling of metals to surface soil horizons by tolerant plants to induce toxic effects on soil flora or fauna.

The scientific literature contains most information on Categories 1 and 2 above, with food-chain contamination and phytotoxicity being widely studied for Cd, Cu, Ni, Pb, and Zn and less information available on other metals.

Some metals pose little hazard to food-chain contamination because of their strong phytotoxic effects, that is, increasing metal concentrations cause mortality before transfer to the next trophic level has an opportunity to occur. Chaney (1980) has termed this the "soil–plant barrier" and classified metals into 4 groups based on

Bioavailability of Metals in Terrestrial Ecosystems: Importance of Partitioning for Bioavailability to Invertebrates, Microbes, and Plants. Herbert E. Allen, editor. © 2002 Society of Environmental Toxicology and Chemistry (SETAC). ISBN 1-880611-46-5

metal retention in soil and metal translocation within the plant. Group 1 is comprised of the elements Ag, Cr, Sn, Ti, Y, and Zr, which pose little risk because they are not taken up to any extent by plants, owing to their low solubility and strong retention in soil (i.e., low concentrations in soil solution) and consequently negligible uptake and translocation by plants. The presence of elevated concentrations of these elements in plants usually indicates direct contamination by soil or dust. Group 2 includes the elements As, Hg, and Pb, which are strongly sorbed by soil colloids, and while plant roots may absorb them, they generally are not readily translocated to aboveground tissues and therefore pose minimal risks to the human food chain. However, they could affect subterranean fauna that graze on the tissue. In terms of total metal concentrations in soil, these metals generally do not raise concerns about phytotoxicity because concerns about their ingestion through dust or soil by humans and animals are triggered at lower concentrations. However, at sufficiently high concentrations, these metals are strongly phytotoxic. Group 3 comprises the elements B, Cu, Mn, Ni, and Zn, which are readily taken up by plants and are phytotoxic at concentrations that pose little risk to human health. Conceptually, the soil–plant barrier protects the food chain for these elements, but hazards from phytotoxicity are more likely to occur. Group 4 consists of Cd, Co, Mo, and Se, which pose human or animal health risks at plant tissue concentrations that are not generally phytotoxic. These metals are of most concern because of bioaccumulation through the soil–plant–animal food chain.

When classifying metals for hazard in this way, it is important to consider the basis on which toxicity is being assessed. If it is being assessed on a nutrient solution or soil solution basis, then rankings may be very different from those determined on a soil concentration or metal loading basis. The latter needs to consider the soil reactions that mitigate toxicity.

The Plant Root and Basic Metal Uptake Mechanisms

Our knowledge of the mechanism of metal uptake by plants comes from studies of metal micronutrient (especially Cu and Zn) uptake by plants, and models are available to describe the process, with varying success (Nye and Tinker 1977; Barber 1995).

The key processes controlling the kinetics of metal uptake by plants are

- passive movement of metal into the root, which does not require energy;
- passive movement of metals in response to an electrochemical gradient established by energy; and
- active metal uptake against an electrochemical gradient, requiring energy (this occurs only if the metal ion has a net negative charge).

The barrier to uptake that requires the use of energy is the plasma membrane inside the cell wall (Barber 1995). Ions move from the soil solution to the xylem vessels in the root by 2 principal processes: apoplasmic or symplasmic flow. Apoplasmic flow occurs when ions do not cross the plasma membrane into the cytoplasm of root cells, but move through the space between cell walls (the root "free space"). This pathway is limited by the Casparian strip in the root (a band of hydrophobic material), which separates the root cortex from the stele (Figure 3-1), and the Casparian strip requires that the ions cross the plasma membrane to permeate this barrier to apoplasmic movement.

The second pathway of metal movement from the cortex to the stele is through the symplasm (Figure 3-1). Metals cross the plasma membrane to the cytoplasm of cells in the cortex and move through the plasmodesmata past the Casparian strip to the stele (Figure 3-1).

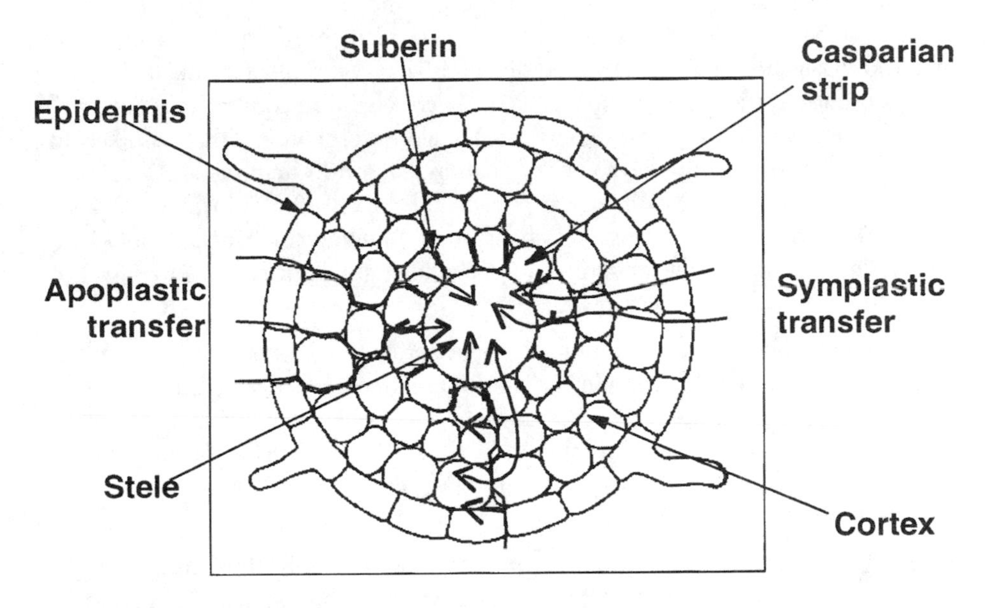

Figure 3-1 Simplified structure of the root, showing pathways of metal transport radially across the root to the stele

Movement of metals in the apoplast is restricted because of the highly tortuous nature of the pores in the cell walls. The actual distance an ion must travel is much greater than the radial distance from the epidermis to the Casparian strip (Clarkson 1993). Furthermore, the walls of these pores are highly (negatively) charged because of the exposed carboxyl groups of the galacturonic and glucuronic acids present in the pectins and hemicelluloses (Clarkson 1988). This provides a strong diffusive barrier to cation movement, akin to the diffusive barrier in soils provided by the permanent and variable negative charge on soil. This large pool of extracellular binding sites for metal also means that, to measure true root uptake across the plasma membrane, it is necessary to remove this extracellularly bound metal from the apoplast; otherwise, plant metal uptake is overestimated. For example, depending on the exposure and growth period, more than 90% of the Zn found in plant roots may be in the apoplasm (Reid et al. 1996).

Metal transport across the plasma membrane to facilitate symplasmic flow to the stele is still only poorly understood. Kochian (1993) and Welch and Norvell (1999) have discussed possible mechanisms for Zn and Cd, respectively. For all cationic metal species, the driving force for uptake across the plasma membrane is the large negative electrochemical potential produced as a result of the membrane H^+ translocating adenosine triphosphatase (ATPase). This potential is of the order of –100 to –200 mV, so that uptake of cationic metals need not be directly linked to a metabolic pathway, but indirectly through the energy required to maintain the potential difference across the membrane through the proton pump. For example, Welch and Norvell (1999) calculate that, given a transmembrane potential of –120 mV and a Cd concentration in soil solution of 1.0 nM, Cd could accumulate in the cell to a concentration greater than 11 μM at electrochemical equilibrium. The actual site of membrane transport is contentious. Some workers contend that it is likely to be an ion channel, and there is some speculation that for divalent metals this may be a Ca^{2+} or Mg^{2+} channel (Kochian 1993; Welch and Norvell 1999). Kochian (1993) suggests that the ferric reductase, which facilitates Fe uptake in dicotyledonous plants, may be responsible for gating of ion channels and hence for facilitating divalent metal ion uptake (Figure 3-2).

A second possible mechanism for metal uptake across the plasma membrane is transport of metal-chelate complexes. In response to metal deficiency, the plant produces and releases chelating agents into the rhizosphere (phytometallophores), which can complex metal ions in solution (Kochian 1993). The complexed metal form is then transported into the plant through a transport protein specific for that phytometallophore (Figure 3-3).

There is certainly evidence that this occurs, as von Wiren et al. (1996) recently demonstrated the cotransport of ^{14}C-labeled phytometallophore with ^{65}Zn into maize (*Zea mays* L.) roots. The rate of metal uptake into plants has usually been described using Michaelis-Menten kinetics developed for describing enzyme reaction kinetics (Figure 3-4).

In the model $F = F_{max}\, c/(K_m + c)$, when $F = F_{max}/2$, $K_m = c$, F is the flux rate of metal and c is the metal concentration in solution. F_{max} is the maximum rate of uptake for the metal, while the parameter K_m describes the affinity of the transporter for that metal. This model often is used as a component of plant uptake models (discussed in "Modeling Metal Uptake by Plants," p 45) to predict ion uptake by plants in soils. Normally, it is assumed that the rate of uptake decreases as metal concentration increases because of the saturation of high-affinity uptake systems. However, at high metal concentration in solution, damage to membranes and transport proteins also may cause a reduction in metal uptake rates, so that low-affinity systems may be confused with the onset of phytotoxicity. Metal uptake experiments usually are

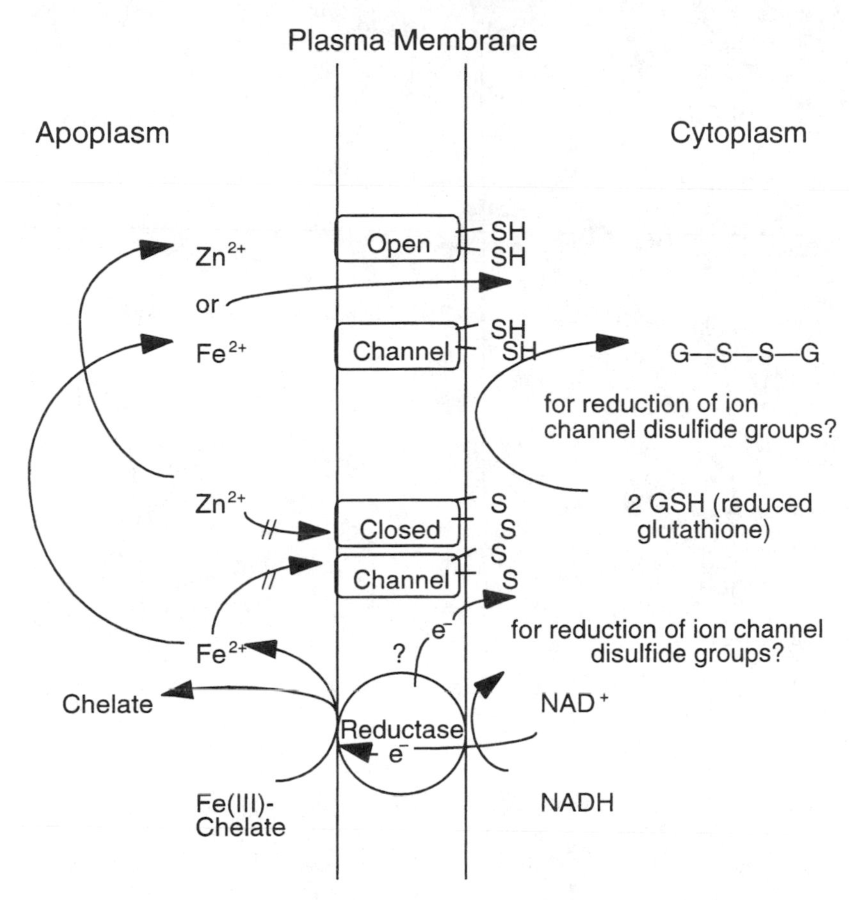

Figure 3-2 Suggested model for Zn^{3+} uptake into root cells of dicots and non-graminaceous monocots. In response to deficiency of Cu, Fe, or Zn, the plasma membrane reductase (that functions in Fe^{3+} reduction and uptake) is stimulated, resulting in the reduction of sulfhydryl groups involved in gating a divalent cation channel that mediated Zn^{2+} influx. (NAD^+ = nicotinamide adenine dinucleotide; NADH = reduced form of NAD^+. Reprinted with permission from Zinc in soils and plants, 1993, p 45–57, Zinc absorption from hydroponic solutions by plant roots, Kochian LV, figure 2, copyright Kluwer Academic Publishers.)

performed over short time periods (<48 h), so that visual symptoms of phytotoxicity may not be apparent. For example, Hart et al. (1998) recently studied the uptake of Cd into wheat (*Triticum aestivum* L.) roots at concentrations ranging up to 1 µM, well above those found in even polluted agricultural soils (50 to 250 nM) and likely to cause damage to root cell components. It is difficult to interpret such data in terms of uptake mechanisms when damage to transport systems is likely.

Many of the mechanisms developed to describe metal uptake by plants are related more to micronutrient deficiency conditions in soil. Mechanisms of uptake at high solution concentrations, indicative of polluted soils, have not been widely studied. At high solution metal concentrations, metal uptake may be controlled more by

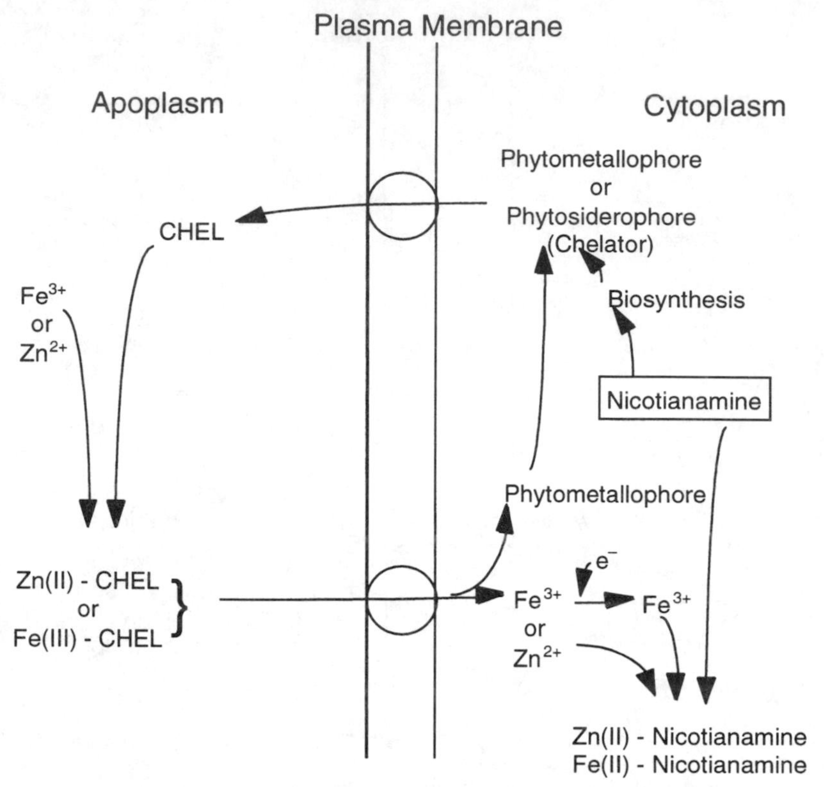

Figure 3-3 Suggested model for phytosiderophore (or phytometallophore)-based uptake of Zn^{2+} into root cells. In response to either Fe or Zn deficiency, phytometallophore release is stimulated in the rhizosphere. These chelators (CHEL) complex and mobilize Fe and Zn, which are then transported by the same transport protein. (Reprinted with permission from Zinc in soils and plants, 1993, p 45–57, Zinc absorption from hydroponic solutions by plant roots, Kochian LV, figure 5, copyright Kluwer Academic Publishers.)

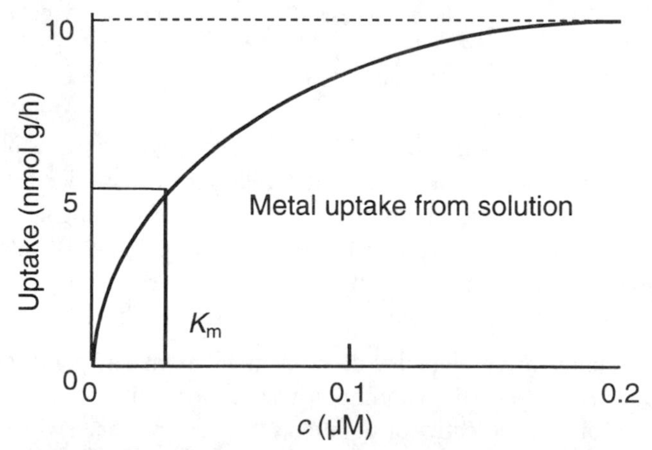

Figure 3-4 Michaelis-Menten kinetics of ion uptake by plant roots

mechanisms regulating metal efflux from the cell than by metal influx. For example, Santa-Maria and Cogliatti (1998) recently showed that efflux of Zn from wheat roots was an extremely important process in the control of Zn accumulation at high external concentrations (up to 1 mM; Table 3-1). Efflux has also been suggested as a mechanism for plant tolerance to high Cd concentrations in solution (Costa and Morel 1993). This is discussed further in the next section.

Table 3-1 Effect of solution Zn concentration on influx and efflux of Zn from wheat roots[a]

Solution Zn concentration (μM)	Zn influx (μmol/g · h)	Zn efflux (μmol/g · h)	Efflux/influx (μmol/g · h)
0.5	0.25	0.18	0.72
10.0	1.75	1.31	0.75
100.0	11.03	9.40	0.85
1000.0	300.43	298.91	0.99

[a] Data from Santa-Maria and Cogliatti 1998.

Modeling Metal Uptake by Plants

Models are a useful way to examine how plant uptake processes, as described by the uptake rates determined in nutrient solution, interact with soil factors controlling the supply of metal to the plant. Models are also a useful way to predict how changes in environmental or soil conditions could affect metal uptake by plants. Several texts are available on this topic (Nye and Tinker 1977; Barber 1995), and only a brief overview is presented here.

The plant root can be regarded as a selective sink for ions in the soil solution. The basic mechanisms by which ions move toward the plant root in soil are discussed fully by Brewster and Tinker (1970) and Barber (1995). Metals may enter the plant root (including root hairs) through the following pathways:

1) interception of solid-phase metal by plant roots ("contact exchange");
2) uptake of metal from soil solution, transported to the root by mass flow of water to the root to replace transpirational losses;
3) uptake of metal from soil solution, transported to the root by diffusion of metal through the solution in response to a concentration gradient induced by selective uptake of metal ion by the root; and
4) uptake of metal by symbiotic microorganisms associated with the root and transfer of metal from the symbiont to the root.

The amounts of metal taken up through contact exchange are regarded to be small, compared to those supplied by mass flow and diffusional processes (Barber 1995), except for alkali earth metals (Ca, Mg).

Mass flow of solution to the root takes place as a result of root absorption of water to replace transpirational losses. Flux of metal to the root (J_r) is therefore a product of the volume of water absorbed (v_o) and the concentration of metal in the soil solution (C_l):

$$J_r = v_o C_l \qquad \text{(Equation 3-1)},$$

where metal uptake by the plant roots exceeds the supply brought to the root by mass flow, a depletion of metal close to the root occurs and metals may diffuse towards the root in response to this concentration gradient. Flux of metal to the root therefore becomes the summation of diffusional and mass flow fluxes:

$$J_r = D_e \frac{\partial C_s}{\partial r} + v_o C_l \qquad \text{(Equation 3-2)},$$

where D_e is the effective diffusion coefficient of the metal in soil, C_s is the concentration of metal on the solid phase that equilibrates with C_l, and r is the radial distance. Barber (1995) gives further discussion of this model and the mathematical solution to the terms.

It is interesting to estimate the relative contribution of mass flow and diffusional fluxes of elements to the plant root. For nutrient elements, (e.g., P, Zn) root uptake often exceeds supply by mass flow so that diffusional transport to the root is significant. Autoradiography has been used to visualize depletion patterns for these elements around actively growing roots. For metals, autoradiographic evidence is restricted to Zn at low levels in soil solution (2.3 to 2.6 μM) (Wilkinson et al. 1968). Model calculations by McLaughlin, Smolders et al. (1998) for Cd at low levels of contamination in soil suggested that some depletion of Cd may occur at the root surface (Figure 3-5), thus making diffusion an important transport process.

This is related to the very high transfer factor between Cd in solution and Cd in plants, which creates a concentration gradient in the rhizosphere despite Cd being relatively weakly sorbed to soil compared to Zn. Cadmium uptake by plants has been shown to have very low K_m values, ranging from 20 to 80 nM (Cataldo et al. 1983; Mullins and Sommers 1986; Costa and Morel 1993; Hart et al. 1998), which is 2 orders of magnitude lower than those for Zn (Chaudhry and Longeragan 1972a; Mullins and Sommers 1986).

In metal-polluted soils, mass flow of metals to the root may deliver a significant amount of metal to the root because of the much higher concentrations in soil solution and the lower rate of uptake compared to macronutrient elements. Some simple calculations can be used to illustrate this point. Water use by crops depends on growth and climatic conditions, but a figure of 250 L/kg can be used as an average (Barber 1995). Concentrations of metals in soil solution vary widely, and most modeling for metals has been undertaken for micronutrients in agricultural soils where solution concentrations are extremely low. In polluted soils, solution metal concentrations may be high; for example, Lorenz et al. (1997) observed levels

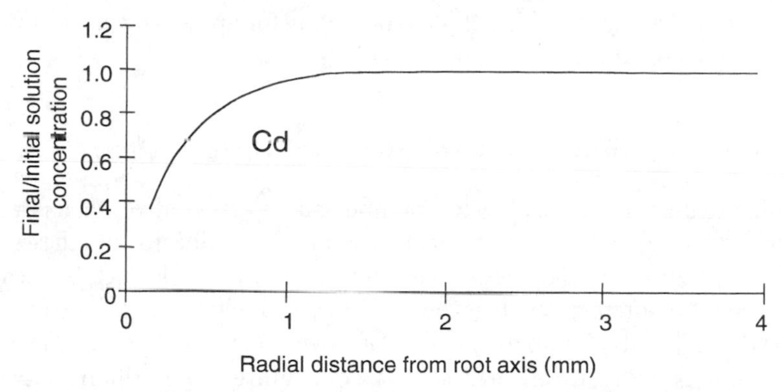

Figure 3-5 Simulated concentration profile of Cd around a spinach root growing in a loamy sand after 3 days of uptake. Solute transport calculation and parameters based on volumetric moisture content, 0.22; root density, 2/cm^2; root radius, 0.02 cm; water flux at root surface, 7.4×10^{-7} cm/s; effective diffusion coefficient in soil, 8.6×10^{-9} cm^2/s; buffer power in soil, 34; initial concentration, 60 nM; root uptake as Michaelis-Menten kinetics per root area with V_{max}, 0.39 nmol/m^2·s, K_m, 76 nM with no lower concentration limit to uptake from solution (Reprinted with permission from McLaughlin MJ, Smolders E, Merckx R. 1998. Soil–root interface: Physicochemical processes. In: Huang PM, editor. Soil chemistry and ecosystem health. Madison WI: Soil Science Society of America. p 233–277. Copyright Soil Science Society of America.)

up to 160 nM for Cd and 55 μM for Zn. These concentrations are sufficient to accumulate Cd and Zn in plant tissue to concentrations of 40 μmol/kg (4.5 mg/kg) and 14 mmol/kg (900 mg/kg), respectively. These concentrations are approaching phytotoxic thresholds for tissue Cd and exceeding those for Zn (Beckett and Davis 1977; Chang et al. 1992), indicating that mass flow may supply sufficient metal in contaminated soils to easily exceed that taken up by the root. In such situations, Zn may even accumulate at the root surface, with a diffusion gradient established away from the root. This is purely speculative, however, and requires experimental validation. Some modeling results provide support for this hypothesis. Barber (1995) presented data for modeled Zn uptake by maize from soil when soil Zn concentrations were raised from 1.0 to 250 mg/kg. Concentrations of Zn in the soil solution at the root surface were predicted to increase above those in the bulk soil by up to 1.25-fold when soil Zn was greater than 150 mg/kg.

Knowledge of the contribution of mass flow and diffusion to metal uptake by plants in polluted soils is critical to developing strategies to minimize metal uptake. As pointed out by McLaughlin et al. (1998b), if metal transport to the root is controlled by diffusion, it can be speculated that uptake can be reduced by factors reducing the effective root surface area of the plant, the affinity of metal uptake transporters (e.g., cultivar differences), and the diffusive mobility of the contaminant through soil (e.g., a reduction in chelating agents or root exudates). On the other hand, if mass flow predominates, the root uptake characteristics (e.g., rate of uptake per unit root length) are rate limiting, and it can be speculated that those factors reducing the uptake kinetics (e.g., addition of ions competing for uptake) are most appropriate to reduce soil–plant transfer of the metal.

Environmental Control of Metal Uptake Processes

In addition to the soil and plant factors outlined above that control metal uptake from soils, plants also respond to other environmental conditions, and these responses may affect metal uptake or metal toxicity. For example, metals reach the root through root interception, mass flow of soil solution to the root, and diffusion of metal through soil solution to the root. Plants will physically access more metal in soil when root growth rates are high, so that temperature and nutrition may affect plant metal uptake directly through effects on plant growth (e.g., Haghiri 1974; Williams and David 1977; Giordano et al. 1979).

Similarly, environmental conditions that affect mass flow to the root, such as temperature and humidity, may affect metal uptake through the rate of metal supply to the root. For example, Blaylock et al. (1997) showed that humidity, and hence transpiration rates, markedly affected uptake of Pb (complexed by ethylenediaminetetracetic acid [EDTA]) by Indian mustard (*Brassica juncea* L.), and Salt et al. (1995) demonstrated that Cd accumulation by the same plant from nutrient solution could be blocked by addition of a stomatal closing agent (abscisic acid), thereby effectively stopping transpiration (Figure 3-6).

Some metals, for example, Zn, appear to be less affected by transpiration rates than Cd. Wilkinson et al. (1968) grew wheat plants at 2 different relative humidities (50% and 95%) and examined Zn uptake by the plants. Despite plant transpiration changing almost 4-fold between treatments, uptake of Zn was unchanged, indicating a close control of Zn absorption rates by the roots. Such homeostasis for an essential element like Zn is not unexpected.

Evidently, crops or climates with high transpiration rates, for example, hot dry climates, will have increased mass flow of solute to the roots than cooler humid climates and are therefore likely to lead to greater transport of metals to the root surface. For elements that appear to be transported indiscriminately to shoots through the xylem, for example, Cd, rates of metal uptake by plants will be increased as the plant grows. Provided that this rate of increase exceeds the shoot growth rate, metal concentrations in leaves will increase as the plant grows (Figure 3-7) and will be greater in hotter conditions.

Ion Interaction in Plant Metal Uptake Processes

As indicated above, metal ion transport across the plasma membrane may be mediated by a number of processes, including transport across divalent ion channels. There is evidence that metal transporters are not highly specific, as several

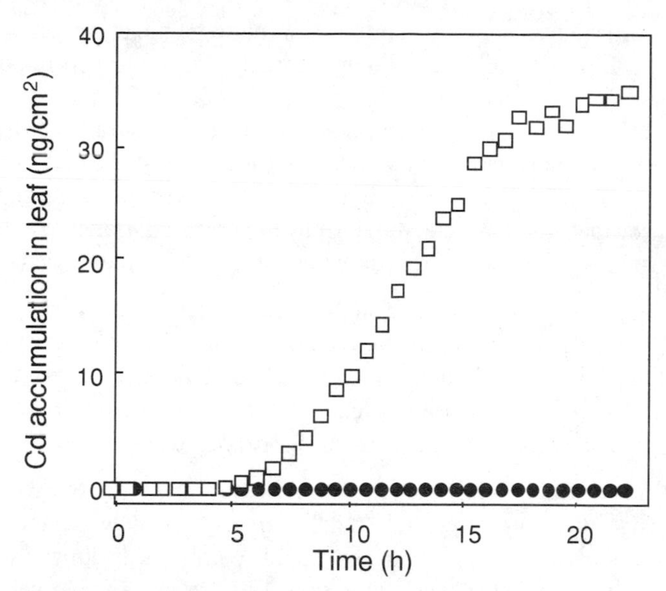

Figure 3-6 Accumulation of Cd in an intact leaf of Indian mustard exposed to 890 nM Cd (0.1 μg/mL) in nutrient solution using plants previously exposed to a stomatal closure agent (●) or control (□) plants (Reprinted with permission from Mechanisms of cadmium mobility and accumulation in Indian mustard. Salt DE, Prince RC, Pickering IJ, Raskin I. *Plant Physiol* 109:1432. 1995. Copyright American Society of Plant Physiologists.)

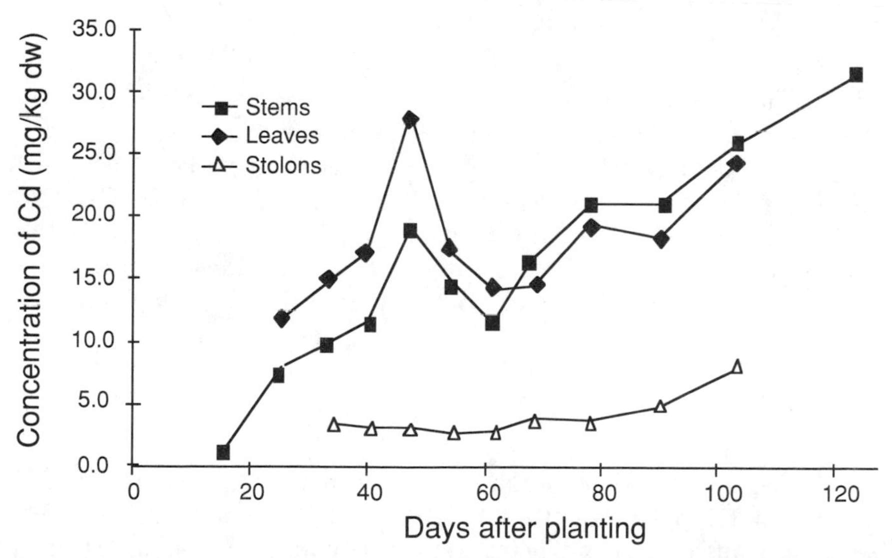

Figure 3-7 Cadmium concentrations in leaves and stems of potato (*Solanum tuberosum* L.) in relation to stage of growth. Plants were grown under field conditions and irrigated throughout the growth period (dw = dry weight; McLaughlin et al. unpublished data).

studies have indicated a competitive interaction between metals in terms of root uptake (Bowen 1969; Chaudhry and Longeragan 1972b, 1972c; Giordano et al. 1974; Iwai et al. 1975; Jarvis et al. 1976; Tyler and McBride 1982; Cataldo et al. 1983; Reid et al. 1996; Smolders, Lambrechts et al. 1997; Smolders, Van den Brande et al. 1997). Given the similarities of hydrated ion size for some of the metals and competing ions (e.g., Cu^{2+} and Zn^{2+}, Cd^{2+} and Ca^{2+}), this is not surprising. Alkaline earth cation competition for metal uptake may affect the relationship between free metal ion activities in solution and toxic effects of metals on plants (Parker and Pedler 1997b).

Protons (H^{+}) may also be important competitors for metal uptake by roots, and in contrast to the effect in which metal availability in soil increases as pH decreases, plant uptake of metal from solution is reduced as pH decreases (Bowen 1969; Chaudhry and Loneragan 1972c; Tyler and McBride 1982; Hatch et al. 1988).

As well as direct competitive effects for membrane transport, there have also been suggestions that micronutrient metal Zn deficiencies may allow greater uptake of contaminant metals, for example, Cd, because of membrane leakiness under conditions of severe Zn deficiency (Oliver et al. 1994; Welch and Norvell 1999).

Ion Speciation and Ion Uptake: Evidence from Nutrient Solution

The Free Ion Activity Model (FIAM) in environmental toxicology (Lund 1990) states that the toxicity or bioavailability of a metal is related to the activity of the free aquo ion. The FIAM is widely used in aquatic toxicology (Campbell 1995) and is gaining popularity in studies of soil–plant relationships (Parker et al. 1995).

Evidence that the free metal ion activity controls plant uptake of metals from solution derives from the work of DeKock (1956) and DeKock and Mitchell (1957), who raised several very important hypotheses with regard to uptake of metals by plants. These authors added nitrilotriacetic acid (NTA), EDTA, and diethylenetriaminepentaacetate (DTPA) to nutrient solutions and found toxicity of Cu, Co, Ni, and Zn to be markedly reduced in line with a reduction in leaf metal concentrations. They were the first to suggest that the charge of the metal-chelate complex could be important in determining plant availability. Several later solution culture studies confirmed DeKock's early suggestion that chelation reduces metal uptake or toxicity (Tiffin et al. 1960; Halvorson and Lindsay 1977; Harrison et al. 1984; Taylor and Foy 1985).

However, there is also evidence that the FIAM may not be valid in all situations. Taylor and Foy (1985), Checkai et al. (1987), Bell et al. (1991), and Laurie et al. (1991a) noted differences in plant uptake of metals at the same metal activity in solution when different chelators were used in nutrient solution. Taylor and Foy (1985) found differential toxicities in wheat exposed to $CuSO_4$ or Cu-EDTA, with the latter producing higher concentrations of Cu in the plant for the same growth reduction. Parker et al. (1992) also noted anomalously high Zn concentrations in apparently Zn-deficient tomato plants (*Lycopersicon esculentum* L.) grown in DTPA-buffered solutions, and they speculated that Zn was present in the leaves as Zn-DTPA complexes. Checkai et al. (1987) examined metal uptake by tomato in resin-buffered solution culture where metal activities were held constant by the chelating resin. Addition of EDTA to these solutions increased markedly the total concentration of metal in solution (the increase being metal-EDTA complexes), while concentrations of Cd and Zn in plant shoot tissues were unaffected. Concentrations of Cu in plant shoots were enhanced in EDTA treatments (approximately 38%), but this is small in relation to the 10^4-fold increase in Cu concentrations in solution. These data suggest that the free metal ions were certainly preferred over the metal-EDTA complexes, but that EDTA affected Cu uptake either through uptake of the intact chelate or better buffering of free Cu^{2+} activities at the site of uptake, leading to enhanced Cu levels in the plant. This latter hypothesis requires the assumption that there is a large diffusive limitation to Cu uptake in the unstirred layer adjacent to the root or in the root apoplast.

Evidence that charge of the metal-chelate complex is important in determining plant uptake can be found in both terrestrial (Sinnaeve et al. 1983; Iwasaki and Takahashi 1989) and aquatic environments (Phinney and Bruland 1994; Vercauteren and Blust 1996). It is now accepted that uncharged metal complexes may exhibit lipophilicity and be transported across biological membranes, and Campbell (1995) documents several examples of this phenomenon from the aquatic literature. Recently, McLaughlin, Andrew et al. (1998) presented data that suggest uncharged $CdSO_4^0$ complexes are equally available to terrestrial plants as the free Cd^{2+} ion. Sinnaeve et al. (1983) showed that uptake of Cu by maize was reduced by addition of Tetren (a chelate that produces a cationic metal complex) but only by 1 order of magnitude for a decrease in free Cu^{2+} activity in solution of 13 orders of magnitude. Translocation of Cu to shoots was enhanced by complexation with Tetren. Similarly, Iwasaki and Takahashi (1989) examined uptake by Italian ryegrass (*Lolium multiflorum* L.) and clover (*Trifolium pratense* L.) of Cu complexed by N, N-EDTA, and Trien. These provided Cu as a divalent anionic complex, an uncharged complex, and a positively charged divalent complex, respectively. Root uptake was reduced by all ligands, but shoot concentrations were reduced by N, N-ethylene-diamine-diacetic acid (EDDA) and EDTA and enhanced by Trien. Enhanced shoot Cu concentrations with Trien were not accompanied by a phytotoxic response. Apoplastically bound Cu was not separated from true root uptake in these experiments and those of Sinnaeve et al. (1983), so that much of the reduction in root Cu uptake in the presence of the chelates was likely due to a reduction in Cu adsorbed in the apoplasm. The data of Iwasaki and Takahashi (1989) indicated that plant uptake of Cu in the presence of Cu-EDDA was greater than that for Cu-EDTA. McLaughlin, Smolders et al. (1997) showed that uptake of Cd and Zn in the presence of a range of chelators producing complexes with negative charges, no charges, and positive charges did not conform to the FIAM (Figure 3-8). The ligands in Figure 3-8 are NTA, EDTA, *trans*-1,2-cyclohexyl-diamine-N,N,N',N'-tetraacetate (CDTA), DTPA, N-2-hydroxyethyl-ethylenediamine-N,N',N'-triacetate (HEDTA), ethylene-bis-(oxyethylenenitrilo)-tetraacetate (EGTA), hydroxyethyl-imino-diacetate (HEIDA), EDDA, 1,4,7,10,13-pentaazatridecane (Tetren), and 8-hydroxyquinoline-5-sulfonate (Sulfoxine).

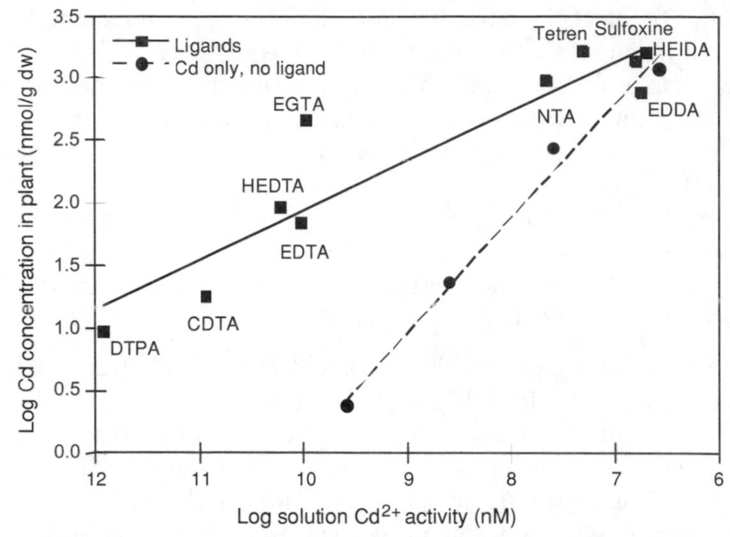

Figure 3-8 Relationship between plant Cd concentration and the negative logarithm of solution Cd^{2+} activity in molar units (pCd^{2+}) in the absence and presence of ligands. Standard deviations of mean values are within the area of the points on the graph. Fitted lines are $y = 5.9 - 0.4x$: $R^2 = 0.87$, $P < 0.001$ (ligands) and $y = 9.22 - 0.92x$: $R^2 = 0.99$, $P < 0.001$ (Cd only, no ligand). (Reprinted with permission from Plant nutrition for sustainable food production and environment, 1997, p 113–118, Plant uptake of cadmium and zinc in chelator-buffered nutrient solution depends on ligand type, McLaughlin MJ, Smolders E, Merckx R, Maes A. Copyright Kluwer Academic Publishers.)

Uptake of Cd was more efficient per unit of free metal in solution as the strength of the Cd-chelate formation constant increased (Figure 3-9).

Reversal of the charge on metal ions by large polydentate ligands has been used in plant physiological studies to trace water movement in plants (Crowdy and Tanton 1970; Tanton and Crowdy 1971). In this work, Pb complexed by EDTA was used as a tracer for water movement in the apoplast of wheat. The EDTA allowed Pb to move freely in the apoplast of the roots, and precipitating the Pb as Pb sulfide made visualization of water movement straightforward through the use of electron microscopy. This is perhaps the best example of the strong diffusional limitation to Pb movement being removed through complexation. It is also interesting to note that Crowdy and Tanton (1970) found that EDTA complexation allowed significant

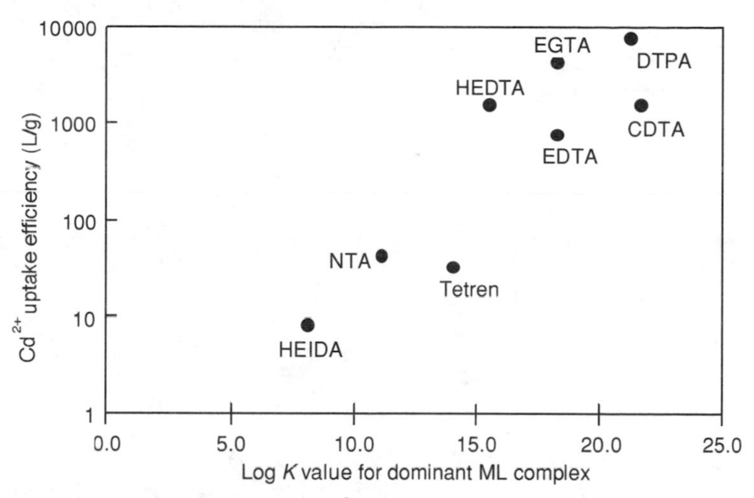

Figure 3-9 Relationship between Cd^{2+} uptake efficiency, calculated as plant Cd concentration per unit solution Cd^{2+} activity, and binding constant of dominant metal–ligand (ML) complex. (Reprinted with permission from Plant nutrition for sustainable food production and environment, 1997, p 113–118, Plant uptake of cadmium and zinc in chelator-buffered nutrient solution depends on ligand type, McLaughlin MJ, Smolders E, Merckx R, Maes A. Copyright Kluwer Academic Publishers.)

translocation of Pb from roots to shoots, approximately 50% of that taken up by the plant. Lead accumulated in the shoot was linearly related to the amount of water transpired, indicating a nondiscriminatory uptake of Pb in the transpiration stream. Later work by Vassil et al. (1998), in relation to phytoremediation of soils, confirmed Crowdy and Tanton's (1970) findings.

Uptake by terrestrial plants of metals complexed by natural organic ligands (humic materials) has not been as widely studied as those complexed by anthropogenic ligands. Tyler and McBride (1982) examined Cd uptake by bush beans and maize from nutrient solutions as affected by addition of Aldrich humic acid (HA). Unfortunately, very high Cd^{2+} activities were employed in this experiment (>500 nM), levels well above those normally encountered in soils (McLaughlin, Tiller et al. 1997), even in contaminated soils (Lorenz et al. 1997). No plant growth data were reported to evaluate phytotoxicity symptoms at these high Cd^{2+} activities. The authors concluded that the free Cd^{2+} ion controlled Cd uptake by roots but did not control Cd accumulation in shoots. It is interesting to note that if Cd uptake in these experiments is plotted against Cd^{2+} activity in solution, uptake at any given Cd^{2+} activity (measured by ion-selective electrode [ISE]) was greater in the presence of HA, compared to non-HA treatments (Figure 3-10).

Cabrera et al. (1988) used high concentrations of Cd (up to 45 μM) in the presence of increasing concentrations of HA (extracted from a soil) to evaluate Cd uptake. In their experiments, HA actually was extremely toxic to the plants in the absence of Cd, but HA certainly mitigated the toxic effects of high Cd through reductions in

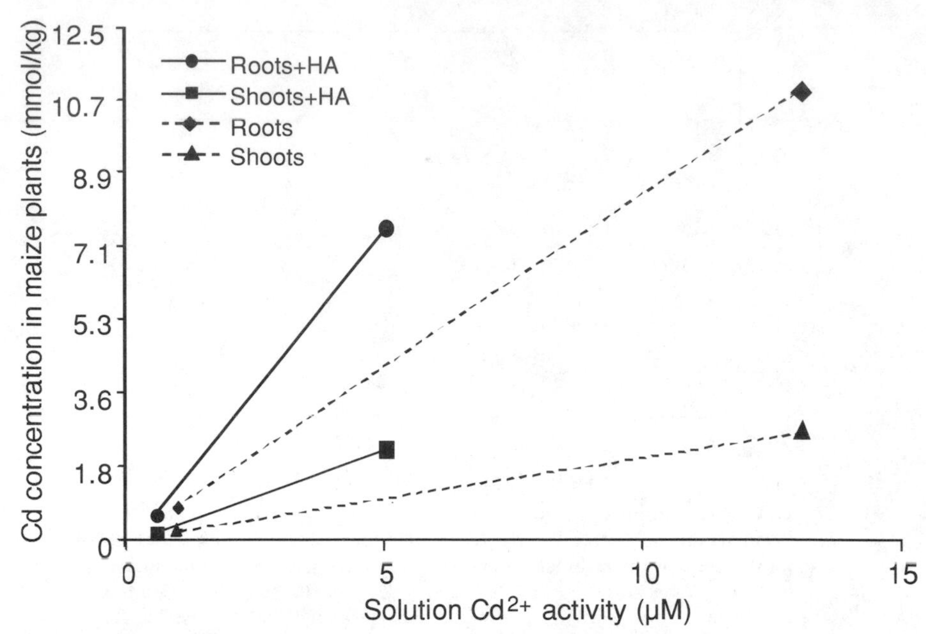

Figure 3-10 Concentrations of Cd in maize roots and shoots in relation to activity of Cd^{2+} in solution, in the presence or absence of HA added at 120 mg/L (replotted from data in Tyler and McBride 1982)

free Cd^{2+} activities (measured by dialysis). However, as for the study of Tyler and McBride (1982), it appeared that plant Cd uptake at a given Cd^{2+} activity was greater in the presence of HA.

There is also evidence that metal complexes with inorganic ligands such as chloride (Cl) may be taken up by plants (Smolders and McLaughlin 1996a, 1996b). In these experiments, which used nutrient solutions both unbuffered and buffered with regard to Cd^{2+} activity, Cd uptake by Swiss chard (*Beta vulgaris*) was found to be related not only to Cd^{2+} activity in solution but also to activities of $CdCl_n^{2-n}$ complexes, formed as a result of increasing Cl concentrations in solution (Figure 3-11).

These data suggest that either the $CdCl_n^{2-n}$ complexes were taken up directly by the root through membrane transport mechanisms or through breaks in the endodermis (Römheld and Marschner 1981) or the presence of labile $CdCl_n^{2-n}$ complexes allowed greater mobility (and hence buffering) of Cd^{2+} in the unstirred liquid layer adjacent to the root, or in the root apoplast, to sites of ion uptake.

Ion Speciation and Ion Uptake: Evidence from Soil-Grown Plants

The link between metal speciation or complexation and metal uptake by plants also has been investigated in soils. This work can be divided into studies that have added

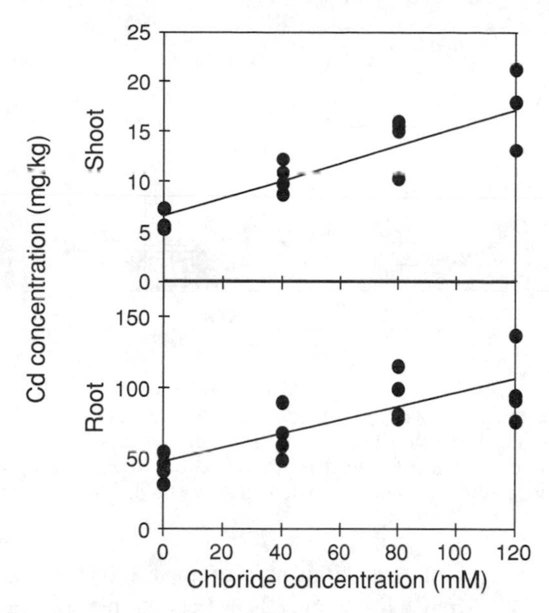

Figure 3-11 Influence of chloride concentrations in solution on uptake of Cd by Swiss chard (*Beta vulgaris*) from solution culture. Solution ionic strength was constant across all treatments (compensated using $Ca(NO_3)_2$), and solution Cd^{2+} activity (4 nM) was held constant across Cl treatments using a chelex resin buffering method. (Reprinted with permission from Smolders E, McLaughlin MJ. 1996. Chloride increases cadmium uptake in Swiss chard in a resin-buffered nutrient solution. *J Soil Sci Soc Amer* 60:1443–1447. Copyright Soil Science Society of America.)

organic or inorganic chelators to soils and determined changes in metal uptake or studies that have investigated natural soils and tried to develop links between measured or calculated metal speciation in soil solution and metal concentrations in plants.

Much of the early work in the former area related to the use of chelates to enhance uptake of Fe from soils, with associated work on metals such as Cd, Cu, Ni, and Zn (Hale and Wallace 1970; Wallace, Romney, Alexander et al. 1977; Wallace, Romney, Cha et al. 1977). In these studies, addition of chelates to soil, usually EDTA or DTPA, were found to enhance metal uptake by plants. For example, Wallace, Romney, Alexander et al. (1977) showed that in a neutral (pH 6) soil contaminated by addition of Cd salts, application of NTA and EDTA significantly increased Cd uptake by bush beans (*Phaseolus vulgaris* L.) (Figure 3-12).

Similar results have been presented more recently to demonstrate chelate-assisted phytoremediation of metal-contaminated soil (Blaylock et al. 1997; Huang et al. 1997; Vassil et al. 1998). Various chelates have been demonstrated to enhance the removal of Pb by plants from contaminated soil, and convincing data have been presented that demonstrate Pb was taken up from soil as a Pb-chelate complex.

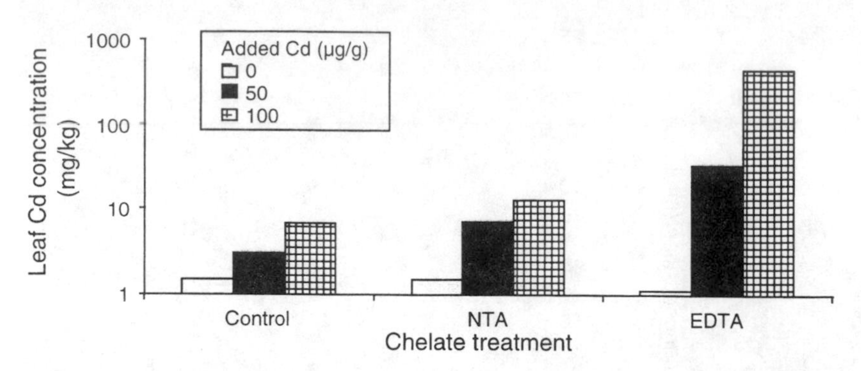

Figure 3-12 Effect of addition of NTA (0.36 mol/g soil) and EDTA (0.27 mol/g soil) to soil contaminated with 2 levels of Cd (as $CdSO_4$) on Cd concentrations in bush bean leaves (data from Wallace, Romney, Alexander et al. 1977). Plant dry weights (data not shown) were unaffected by Cd in the absence of chelates, or by chelates in the absence of Cd. However, when Cd and chelates were combined, phytotoxicity resulted.

However, the high chelate concentrations used in these and earlier studies (Wallace, Romney, Cha et al. 1977) may have damaged the plasma membrane and normal mechanisms for regulating metal transport into the root so that uptake of metal chelates was promoted.

Metal complexation by inorganic ions and the effect on plant metal uptake also has been studied in soil systems (Bingham et al. 1983, 1984; Smolders, Lambrechts et al. 1997). Addition of Cl to soils was found to increase Cd uptake markedly (Bingham et al. 1983, 1984), but the complexation effect of Cl was masked in these early studies by the effect of the counter cation Na. High levels of Cd were added to soils prior to imposing salinity treatments so that Na easily displaced Cd from soil into solution. Thus, interpretation of the results erroneously concluded that $CdCl_n^{2-n}$ complexes were not available for uptake by the plants. Later work by Smolders, Lambrechts et al. (1997), using soils unamended with Cd salts, indicated that Na/Cd exchange is negligible after addition of NaCl to soil and that the increase in Cd concentrations in solution are predominantly $CdCl_n^{2-n}$ complexes in response to complexation by Cl. Increases in plant Cd concentrations in this scenario are due to the uptake of these $CdCl_n^{2-n}$ species, or their ability to buffer Cd^{2+} activities at the plasma membrane.

The second suite of evidence for a link between metal speciation and plant metal uptake derives from experiments in which metal speciation has been determined in a series of soils and related plant metal uptake. This relies on the natural variation in metal complexation in soil solution to be reflected in plant metal concentrations. There are few studies where this has been attempted, mainly because of the difficulties of measuring metal speciation in soil solutions at the concentrations normally encountered in the environment. Cu is perhaps the exception, where the Cu-ISE provides a reliable and convenient way to determine Cu^{2+} activities in solution to levels normally encountered in soils. Minnich et al. (1987) describe an experiment in which Cu levels in soil were varied by addition of $CuSO_4$ and sewage biosolids. Total Cu concentrations in soil solution were only poorly related to plant Cu concentrations, while concentrations of free Cu^{2+} were closely related to plant Cu concentrations. However, only one soil was used in this study, and plant Cu was also very closely related to total Cu added.

A second experiment, described by Sauvé et al. (1996), related Cu speciation in soil extracts and Cu uptake by lettuce (*Lactuca sativa* L.), radish (*Raphanus sativus* L.), and ryegrass (*Lolium perenne* L.) in 8 contaminated soils from urban areas. The authors suggested that a measure of free Cu^{2+} in $CaCl_2$ extracts (pCu^{2+}) was the best predictor of Cu concentrations in the plants. However, there are some uncertainties with regard to interpretation of these data. No plant growth data were given with which to evaluate toxicity of Cu in these soils. It could be argued that any soil test procedure to predict phytotoxicity should use plant shoot or root growth as a variate, rather than plant metal concentration or uptake. There is often a nonlinear relationship between plant metal concentrations (or uptake) and metal supply at low levels of supply for essential metals (see Figure 3-4) and for many metals under toxicity conditions (Davis and Beckett 1978). The regressions were also weighted substantially by data for one soil, which, if removed from the data set, reduced the effectiveness of pCu^{2+} as a predictor of plant uptake, so much that total Cu became equal to or more effective than pCu^{2+}. With this in mind, the data of Sauvé et al. (1996) do not provide strong evidence that consideration of free Cu^{2+} ion in soil extracts considerably improves the relationship with plant uptake.

Lorenz et al. (1997) used 10 contaminated soils to examine the relationship between Cd and Zn in soil solution and accumulation of these elements in leaves and tubers of radish plants. Consideration of metal activities in solution, determined by a resin procedure (Holm et al. 1995), did not improve the relationship between plant and solution metal concentrations, although most of the metal in solution was present as the free ion in several of the soils.

Evidence that complexation of metal affects plant metal uptake is clear for Cd and Cl from several field investigations (Li et al. 1994; McLaughlin et al. 1994; McLaughlin, Tiller, Smart 1997). McLaughlin, Tiller, and Smart (1997) calculated Cd speciation in soil solution for 50 soils, which varied widely in potato (*Solanum tuberosum* L.) Cd concentrations, and found no relationship with calculated free Cd^{2+} in soil solution. Plant Cd concentrations were related to the concentration of all $CdCl_n^{2-n}$ species in solution, in response to Cl added to the soils in irrigation waters. Unpublished data from this study are presented in Figure 3-13 and Table 3-2.

Because of the autocorrelation of the various $CdCl_n^{2-n}$ species activities in solution, it is difficult to separate the actual ionic species contributing to enhanced Cd uptake by the potatoes (Table 3-2). However, it is clear that Cd^{2+} does not provide a good predictor of plant Cd uptake in this situation.

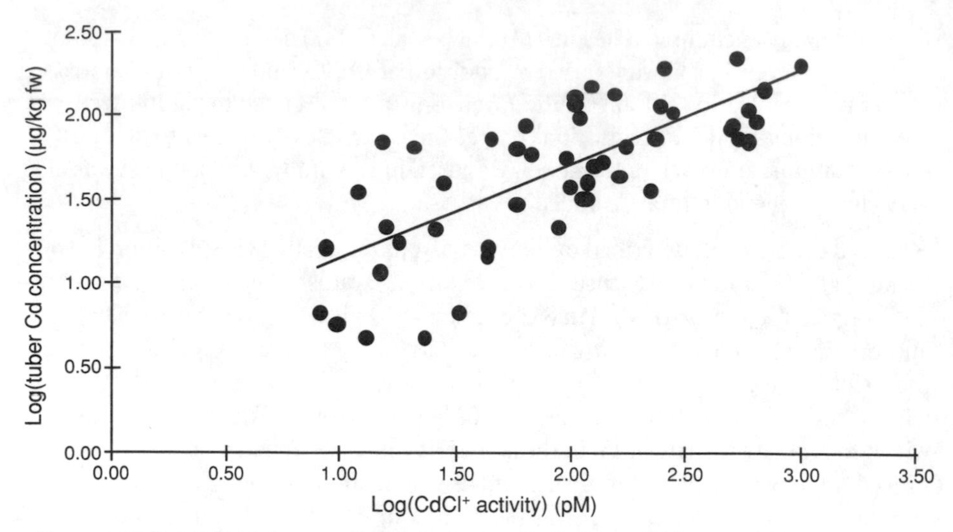

Figure 3-13 Relationship between $CdCl^+$ species in soil solution and Cd concentrations in potato tubers (both variates log transformed) from 50 field sites in South Australia (fw = fresh weight; McLaughlin et al. unpublished data)

Table 3-2 Regression coefficients (R^2) for linear relationships between ionic Cd species in soil solution and Cd concentrations in potato tubers (both variates untransformed and log transformed) from 50 field sites in South Australia[a]

Ion species	Untransformed	Log transformed
Cd_T	0.30	0.47
Cd^{2+}	0.00	0.03
$CdCl^+$	0.40	0.55
$CdCl_2^0$	0.44	0.58
$\Sigma CdCl_n^{2-n}$	0.45	0.57
Cl	0.57	0.47

[a] Source: McLaughlin et al. unpublished data.

It is possible that despite the low activities of $CdCl_2^0$ in solution, the uncharged nature of this ion may lend it lipophilic characteristics. Gutknecht (1981) showed that diffusion of Hg^{2+} through lipid bilayers was markedly enhanced by complexation to form $HgCl_2^0$ (Figure 3-14), so that $CdCl_2^0$ may indeed be the permeating ion responsible for enhanced Cd uptake by plants in the presence of high Cl concentrations.

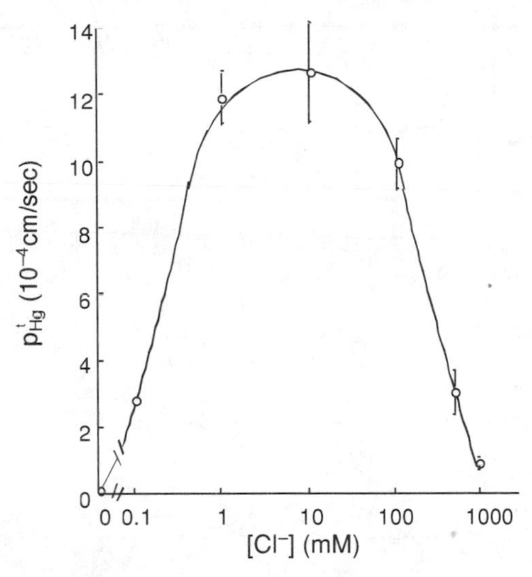

Figure 3-14 Permeability of Hg (P^t_{Hg}) through lipid bilayers as a function of Cl concentration. The membranes were formed from egg lecithin plus cholesterol (1:1 mol ratio) in tetradecane. Donor solutions were $Hg(NO_3)_2$ with NaCl varying from 0.1 to 1.0 M, and receptor solutions were $NaNO_3$ solutions varying from 0.1 to 1.0 M with EDTA (1.0 mM). All solutions were buffered at pH 7.0 with 4-(2-hydroxy-ethyl)-1-piperazineethanesulfonic acid (HEPES). (Reprinted with permission from Inorganic mercury (Hg^{2+}) transport through lipid bilayer membranes. Gutknecht J. *J Membrane Biol* 61:61–66. 1981. Copyright of Springer-Verlag.)

A second hypothesis to explain enhanced metal uptake at constant free metal activity in solution is the presence of diffusional barriers to metal ion transport to uptake sites. In the aquatic literature, Tessier et al. (1994) has termed this the "diffusion layer" and related it to the layer of static solution surrounding the cell (Figure 3-15).

In the terrestrial environment, the diffusion layer for plants may be much thicker than for aquatic organisms because, in addition to the unstirred layer of solution surrounding the root, the metal ion has to diffuse through the mucigel and the apoplast. For plant roots, the unstirred layer is wider than the diameter of an epidermal cell (Clarkson 1988), between 10 and 100 μm. The diffusion length in the apoplast may be greater than 100 μm because of the tortuous nature of the pores in the cell walls (Clarkson 1988). In addition to these plant-based diffusional limitations, the soil itself also exerts a strong diffusional limitation to metal transport to the root, due to the strong nature of metal binding to soil solid surfaces and the tortuous nature of soil pores (Nye and Tinker 1977; Barber 1995). This strong diffusional limitation to metal transport to sites of uptake is the likely reason for chelate-enhanced metal uptake observed in soil-grown plants, especially at low rates of chelate application (DeKock and Mitchell 1957; Hale and Wallace 1970; Wallace, Romney, Alexander et al. 1977; Wallace, Romney, Cha et al. 1977; Huang et al. 1997).

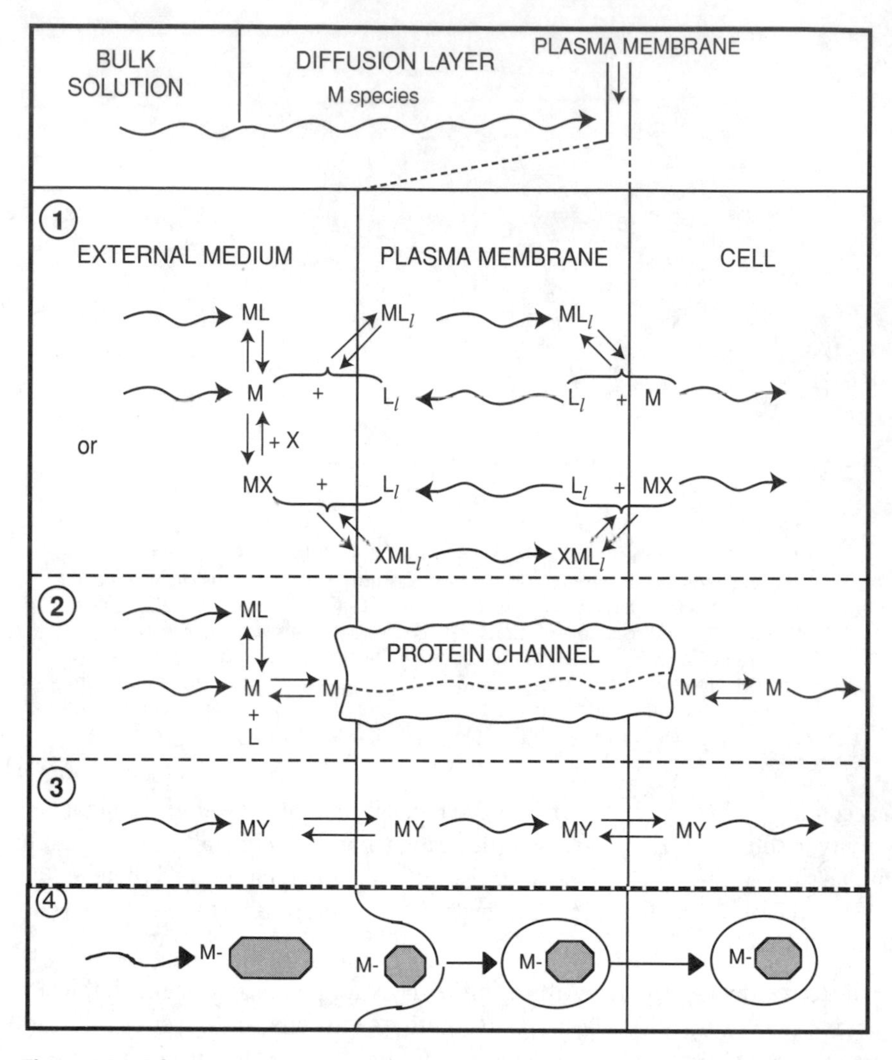

Figure 3-15 Schematic representation of metal transport across the plasma membrane of an aquatic organism. L, X, and Y are ligands. ① = carrier mediated transport, ② = transport through protein channels, ③ = passive diffusion, ④ = endocytosis. (Reprinted with permission from Tessier A, Buffle J, Campbell PGC. 1994. Uptake of trace metals by aquatic organisms. In: Bufle J, De Vitre R, editors. Chemical and biological regulation of aquatic systems. Boca Raton FL: Lewis Publishers. p 197–230. Copyright CRC Press, Boca Raton, FL, © 1992.)

Parker and Pedler (1997b) discussed the importance of root-binding sites in relation to ion complexation in solution and modeled the conditions under which the FIAM would be expected to fail, even where it is assumed that only the free metal ion is membrane transported. Regardless of whether the free ion is taken up by the plant or uptake occurs by some complexed form through direct uptake or alleviation of diffusional limitations, it is evident that the relationship between free metal activity in solution and plant metal uptake is not as close a relationship as previously assumed. Thus, improvements in methodology for speciation of contaminant forms in soil solutions will provide useful information on the potential for transfer of contaminants to the food chain only if parallel information is gathered on the phytoavailability of contaminants in both free and complexed forms.

Rhizosphere Processes: Are They Important?

Plants can modify considerably the chemistry of the soil and soil solution in the rhizosphere, and this topic has been well reviewed in several texts recently (Marschner 1995; McLaughlin, Smolders et al. 1998). McLaughlin, Smolders et al. (1998) list a number of effects exerted by plants on the physicochemistry of soil adjacent to the root:

- Plant uptake may reduce ion activity and desorb metals from surfaces, or convective flow of solution to the root may move additional contaminant to the rhizosphere, leading to sorption.
- Plant-induced changes in solution chemistry can affect sorption, for example, pH, ionic strength, or macronutrient cation concentrations (Ca^{2+}).
- Plants excrete organic ligands, which may increase or decrease the total concentration of metal ions in solution, depending on whether free activity is well buffered or poorly buffered, respectively.
- Living or dead plant material in the rhizosphere can act as new sorbing surfaces for metals.
- Microbial activity, stimulated by plants, can also affect metal behavior by the above processes.

Discussion of these factors will not be repeated here; the reader is referred to McLaughlin, Smolders et al. (1998) for further details of the general concepts.

With regard to plant metal uptake, most studies on rhizosphere chemistry have concentrated on plant strategies to improve micronutrient metal uptake from soil by definition at very low solution activities (Marschner et al. 1986, 1989; Merckx et al. 1986; Morel et al. 1986; Mench et al. 1988; Treeby et al. 1989; Mench and Martin 1991; Mench and Fargues 1994; Gries et al. 1995). Despite the large amount of work spent identifying compounds in the rhizosphere that potentially are able to modify availability of metals to plants, work with isotopic labeling of labile metal pools has suggested that a wide range of plants access the similar pools of Cd and Zn in soils (Hamon et al. 1997) (Figures 3-16 and 3-17).

Less work has been undertaken in polluted soils to determine whether plant modifications to the rhizosphere affect metal uptake and tolerance. Certainly, exudation of organic acids has been implicated in plant tolerance to Al toxicity (Delhaize et al. 1993).

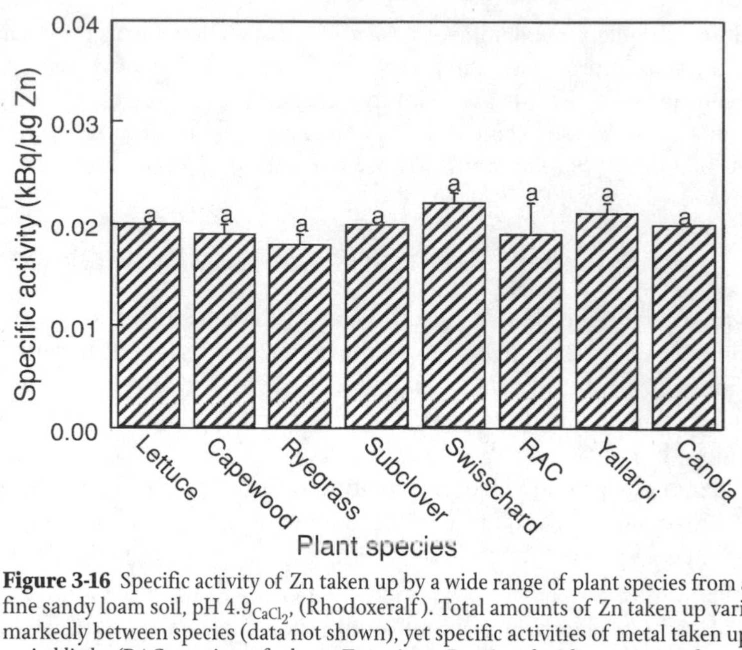

Figure 3-16 Specific activity of Zn taken up by a wide range of plant species from a fine sandy loam soil, pH 4.9_{CaCl_2}, (Rhodoxeralf). Total amounts of Zn taken up varied markedly between species (data not shown), yet specific activities of metal taken up varied little. (RAC = variety of wheat, *T. aestivum*. Reprinted with permission from Hamon RE, Wunke J, McLaughlin MJ, Naidu R. 1997. Availability of zinc and cadmium to different plant species. *Aust J Soil Res* 35:1267–1277. Copyright CSIRO Publishing.)

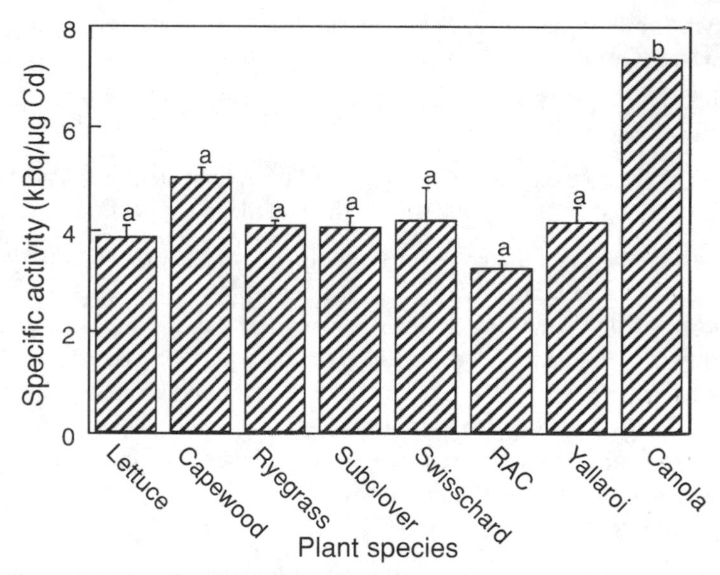

Figure 3-17 Specific activity of Cd taken up by a wide range of plant species from a fine sandy loam soil, pH 4.9_{CaCl_2}, (Rhodoxeralf). Total amounts of Cd taken up varied markedly between species (data not shown), yet specific activities of metal taken up varied little. (Reprinted with permission from Hamon RE, Wunke J, McLaughlin MJ, Naidu R. 1997. Availability of zinc and cadmium to different plant species. *Aust J Soil Res* 35:1267–1277. Copyright CSIRO Publishing.)

Effects of Plant–Microbe Symbiosis on Metal Uptake

The infection of plant roots by ecto- or vesicular arbuscular mycorrhizae may markedly affect plant uptake of metals from soils (McLaughlin et al. 1998b). However, it is important to distinguish the effect of mycorrhizae on metal uptake from soils with low levels of metals from the effect of mycorrhizae in soils that are grossly polluted by metals.

At low levels of metal in soil, infection by mycorrhizae increases uptake of metal contaminants from soil (Cooper and Tinker 1978; Rogers and Williams 1986; Li et al. 1991). Presumably, the mycorrhizae in these situations are acting in a similar way to their enhancement of macronutrient uptake, that is, extension of the effective rooting volume explored by the root–mycelial mass. Hence, metals that are likely to be limited by diffusion through soil to the root (e.g., Cd, Cu, Pb, Zn, Ni) are most likely to be enhanced by mycorrhizal infection.

At high metal concentrations in soil, there is contradictory evidence regarding the effect of mycorrhizal infection on metal uptake by plants. For example, Killham and Firestone (1983) reported that colonization of roots of bunchgrass (*Erharta calycina*) by mycorrhizae enhanced uptake of Cu, Ni, Pb, and Zn and enhanced toxicity of these elements. Other authors have reported that mycorrhizal infection inhibits metal uptake by plants in polluted soils and affords some protection to the plant from metal toxicity (Dueck et al. 1986; Heggo et al. 1990). This has been attributed to a number of factors: binding of metal in the external mycelium or binding to cell walls or membranes in mycelia within the host plant. Metal binding appears to be ecotype related, with mycorrhizae from polluted soils having a greater ability to bind metals than those from noncontaminated soils. Galli et al. (1994) recently reviewed the various factors responsible for binding of metals by mycorrhizal fungi. They concluded that metal binding is predominantly related to sorption to cell wall components such as chitin, melanin, cellulose, and cellulose derivatives, rather than being due to coprecipitation within polyphosphate granules, as previously suggested.

Implications for Predicting Phytoavailable Metals in Soils

There has been extensive study of soil test methods to predict the uptake by plants of essential trace metals from soils in the context of crop nutrition. Assessments of methods to predict metal uptake by plants in contaminated environments are less common. McLaughlin, Maier et al. (1999) suggested a number of criteria that should be applied to any index to diagnose or predict hazards or risks from contamination of soils by Cd, but these apply equally to other metals. The soil test should

1) be relatively simple, inexpensive, and robust;
2) be calibrated under field conditions across a wide range of soil types;
3) be independently validated;

4) account for the major environmental factors known to affect crop metal concentrations or toxic response to plants or organisms (for assessing current hazard or risk, i.e. diagnosis); and
5) (for prognosis) be truly predictive. For example, for crop quality, measurements prior to planting of the crop must be correlated against plant measurements at harvest and not correlated between measurements on soil and plant samples at the same point in time. This is particularly important for Cd in irrigated crops where soil properties, such as pH, effect concentration (EC), and Cl concentrations, can change markedly throughout the growth season (Maier et al. 1997), or where soil cultivation practices by the farmer could change metal distribution and crop rooting patterns in the soil. Where heavy metals are added as part of farm management practices after planting of the crop (and hence after soil sampling for predictive purposes), this must also be taken into account.

Unfortunately, few studies in the literature fulfill these criteria sufficiently to conclude that one procedure is more useful than another in predicting phytoavailable metal in soils.

Early studies were hindered by the inability of analytical instrumentation to accurately determine metals at low concentrations. Researchers often preferred extractants that released a large portion of the solid-phase metals to the extractant solution, for example, chelating agents such as EDTA or DTPA (Lindsay and Norvell 1978) or strong acids such as HNO_3 and HCl. Analytical limits of detection for many metals and metalloids have fallen considerably over the last 20 years because of the introduction and more widespread availability of electrothermal vaporization atomic absorption and inductively coupled plasma atomic emission and mass spectroscopic instrumentation. Weaker extractants such as 0.01 M $CaCl_2$ (Houba et al. 1990) or 1 M NH_4NO_3 (Prüeß 1997) for extracting metals from soils have thus been suggested.

Many of the studies in the literature report concentrations of extractable metals in relation to various soil treatments, for example, liming or other amelioration, sludge–waste (metal) addition, rhizosphere effects or time (Schalscha et al. 1980; Chang et al. 1984; Lake et al. 1984; Van der Watt et al. 1994; Krishnamurti et al. 1995; Berti and Jacobs 1996). While it is valid to use soil extractions to trace changes in metal solubility that result from various treatments, these studies provide little assistance to hazard classification or risk assessment for plants in soils that are contaminated with metals. Indices must be calibrated across a range of soils and environmental endpoints to be useful. As discussed earlier, food-chain contamination and phytotoxicity are the principal risks that must be predicted.

Some studies have attempted to relate metals extracted by various reagents to metal uptake by plants, with much of the early work related to micronutrient nutrition of agricultural crop species. The use of EDTA and DTPA as soil extractants for predicting metal uptake by plants stems from this agricultural heritage (Lindsay and

Norvell 1978). Few studies have specifically addressed the question of predicting the environmental risks from metal contamination of soils. Because recent reviews of this topic have been completed (Lebourg et al. 1996; McLaughlin, Stevens et al. 1999), only the links with plant mechanisms of metal uptake are discussed here.

Given our knowledge of the mechanisms of metal uptake by plants, it is evident that total concentrations of metals in soil is not a useful predictor of risk. Plants access metals in soil principally though the soil solution. At first glance, this suggests that determination of soil solution concentrations (or activities) of metals would provide the best predictor of risk.

However, the limitations of simply considering an equilibrium partitioning approach (and assuming bioavailable metal is that in soil solution) are demonstrated by the relative magnitude of changes in metal uptake by plants and metal partitioning in soil as a function of pH. From Chapter 2, it is clear that soil pH has a major effect on metal partitioning in soil and is the major factor controlling metal concentrations in soil solution. However, effects of liming on metal uptake by plants are often much less pronounced, or may even be the opposite to that predicted by equilibrium partitioning (Andersson and Siman 1991; Maier et al. 1997). It has been suggested that this is due to H^+ competition for uptake sites at low pH or through other ionic interactions affecting uptake processes, for example, Ca^{2+} or other metals (Smolders et al. 1999).

Studies of heavy metal ion concentrations (and speciation) in soil solution in relation to plant uptake are mostly limited to research investigations of soil and plant processes on single (or few) soils, rather than development of new bioavailability indices for a wide range of soils (Bingham et al. 1983, 1986; Hamon et al. 1995; Lorenz et al. 1997). Very few studies examine the link between metal concentrations in soil solution and plant metal uptake or toxicity (Gerritse et al. 1983; Sauvé et al. 1996). Gerritse et al. (1983) examined the relationship between Cd, Pb, and Zn in lettuce, spinach, and potato plants and metals extracted by water ("soil solution") and increasing concentrations of $CaCl_2$ + NaCl + KCl, as well as total metals. They found that relationships between plant metal concentrations and soil parameters were poor with total metals. Better relationships were obtained using extractable metals, although the log-log transformation of the data still masked significant variation in the relationships. Gerriste et al. (1983) found soil-solution Cd concentrations to be useful predictors of tuber Cd concentrations. However, McLaughlin, Tiller et al. (1997), in a study of Cd uptake by potato tubers, found that total concentrations of Cd in solution were not a good predictor of uptake, but the speciation of Cd in solution also needs to be considered if solution Cd measurements are to be useful. This result is perhaps indicative of the problem with using soil-solution metal concentrations alone to predict metal uptake by plants. Where plant uptake causes a depletion of metal in solution, it is highly unlikely that determining metal concentration in solution will provide a good predictor of plant metal concentrations or uptake. Furthermore, plant growth rate drives transpiration and ion uptake rates so that factors affecting plant growth (e.g., soil nutrient status) also may affect the relationship between solution metal concentrations and plant uptake. Indices are also needed that account for the rate of supply of metal to the solution and then to the root, hence the good relationship of plant Cd concentrations with chloro-complexed Cd species in soil solution (or irrigation water salinity; McLaughlin, Tiller et al. 1997; McLaughlin, Maier et al. 1999).

Assessment of the pool of solid-phase metal that buffers the solution metal concentration is another means of accounting for the supply rate term noted above. Hence many researchers have sought to identify soil extraction techniques that measure this pool. In Switzerland and Germany, this has led to the adoption of soil extraction techniques as regulatory standards to control metals in soil. In Switzerland, extraction of metals from soils by 0.1 M $NaNO_3$ is used to assess risks due to metal contamination, and this method appears to be a much better predictor of metal bioavailability than total metal concentrations (Gupta and Aten 1993). This extractant was chosen because the authors suggested that it provided a better predictor of plant metal uptake for several metals compared to other procedures, including total metal concentrations (Table 3-3). The data are not convincing, however, and there is little difference between the dilute salt extractants as predictors of metal uptake. In the Baden-Württemburg region of Germany, 1 M NH_4NO_3 is now used as the standard method of assessing metal bioavailability for regulatory purposes (Prüeß 1997), with this technique now being a recognized Deutsch Institut für Normeng standard (DIN 1995). Correlations between extractable metals in soil and metals in the shoots of several plants were determined on 400 paired soil–plant samples under field conditions using soils and plants from contaminated sites. While there was still considerable scatter in the relationships obtained, the study was well executed, and the author recognized that factors such as pH and other elements interact to affect trace element availability to plants (e.g., Cd and Zn) and included these in the models developed.

Two important points to stress again, however, are that not all metals behave similarly in terms of plant uptake and that there are different pathways for expression of risk, for example, food chain or phytotoxicity. Cadmium contamination of soil is usually at low concentrations compared to Cu, Ni, Pb, or Zn (soil solution concentrations of Cd are usually nM), and risk is principally through food-chain transfer. Plants take up Cd extremely efficiently (Smolders, Van den Brande et al. 1997), and depletion of Cd concentrations in the soil solution in the rhizosphere is likely (but unproven) (McLaughlin, Smolders et al. 1998). Hence, it is probable that any index of Cd bioavailability needs to include factors affecting Cd diffusion through soil to the plant root (e.g., Cl) or an estimation of the ability to replenish soil solution concentrations after plant uptake (e.g., extractable Cd in soil). For the other metals, risk from plant uptake is principally expressed as phytotoxicity, where metal concentrations in soil and soil solution are high (usually μM to mM) and plant uptake may not deplete soil solution concentrations to the same extent (again unproven). Diffusional and desorptive limitations to plant uptake may be less

Table 3-3 Coefficients of determination for the relationship between extractable Cd, Cu, and Zn (log transformed) in selected extraction media and Cd, Cu, and Zn in lettuce and ryegrass for 13 soils in miniplots in the field[a]

	Lettuce			Ryegrass		
Extractant	Cd	Cu	Zn	Cd	Cu	Zn
2.0 M HNO_3	–0.336	0.585	–0.123	0.128	0.894	0.286
0.5 M CH_3COONH_4+EDTA	–0.076	0.603	0.021	0.359	0.891	0.411
0.1 M $NaNO_3$	0.683	0.578	0.861	0.669	0.885	0.752
0.05 M $CaCl_2$	0.676	0.601	0.877	0.787	0.829	0.832
0.1 M $CaCl_2$	0.673	0.598	0.831	0.732	0.867	0.799
0.1 M NH_4NO_3	0.615	0.552	0.854	0.691	0.844	0.686
0.1 M KNO_3	0.603	0.576	0.883	0.584	0.779	0.709

[a] Compiled from data in Gupta and Aten 1993.

important in this situation, so that it is possible that toxicity could be well predicted by consideration of soil solution concentrations or activities. However, no studies have adequately tested this hypothesis. Even in this situation of "demand-driven" uptake, ion competition effects for uptake (e.g., H^+, Ca^{2+}, "Ion Interaction in Plant Metal Uptake Processes," p 48) may confuse the relationship with free metal activity in solution.

Conclusions and Cautions

It would appear that for Cd, Zn, and perhaps Cu, indices of phytoavailability in soil based on determination of the most available metal pools, while not perfect, appear to better predict risk than do total metal concentrations or metals removed from soil by strong extractants. There are insufficient data to determine whether soil solutions or soil extracts are the best indices to use. The additional utility of considering free metal ion activities in soil extracts or soil solution is questionable at this stage and requires further study to resolve. The conditions under which the FIAM fails for plant metal uptake need definition, which will require basic research into mechanisms of metal uptake from solution by plant roots and more information on gradients in metal concentrations in the rhizosphere.

Data for As, Cr, Hg, Ni, Pb, and Se are lacking in terms of development of indices of phytoavailability in soils.

A word of caution is also appropriate. Unlike total concentrations of metal, indices of metal bioavailability can change over time because of changes in the factors that control phytoavailability. For example, soil-to-solution partitioning can change because of changes in soil pH or organic matter content, and solution speciation may change because of the addition or decomposition of organic matter or changes in soil salinity. Depending on how the index of phyto- or bioavailability is used (e.g., screening level, action level, remediation endpoint), it should be recognized that any target value based on a measure of bioavailability is liable to change over time in line with soil conditions.

CHAPTER 4

Bioavailability of Metals to Soil Microbes

Stephen P. McGrath
IACR – Rothamsted

Soil microbes consist of bacteria, actinomycetes, fungi, algae, and protozoa. Because of the large numbers of groups present and their different life histories, it is important to realize that the soil microbial community is very complex. It consists of tens of thousands of species, and it has been estimated that the number of bacterial cells alone in a single gram of soil exceeds the human population of the earth (Skinner et al. 1952). Soil microorganisms constitute less than 0.5% of the soil mass (weight/weight basis), and usually they constitute 2% to 4% of the organic C in soils. Collectively, the soil microbial community is referred to as the "soil microbial biomass" (Jenkinson and Ladd 1981).

Despite their small size, soil microorganisms are essential to soil processes, such as nutrient cycling and biodegradation of organic pollutants, and to the physical properties of soil. Because they cannot be seen, there is a tendency to ignore their importance, but they are vital parts of a healthy ecosystem. Partly because of their small size, diversity, and location in different physical locations in soils, their measurement is difficult. All of these factors impact the availability of data about the effects of metal toxicity on soil microbes. For example, assays for particular species or groups are not simple, unlike those for earthworms, in which particular species can be exposed to a soil, then easily retrieved and their metal uptake and body mass determined after a specific time period. Alternatively, the natural population of earthworms can be extracted from a gradient of polluted soils, and their numbers and metal burden can be established with relative ease. Similar types of measurement of either added or indigenous microorganism populations are not easy to make in soils. Even though soil protozoan assays have been developed, which are analogous to adding earthworms to soil, these suffer from problems because of the size of the organisms and the alterations of the soil during the assay. Because of the small size of protozoa and the consequent difficulties in retrieving them from whole soil samples, the assays generally use soil extracts, and these are amended with "food" substrates to sustain the protozoa in an active state (Bowers et al. 1997; Campbell et al. 1997). Both aspects perturb the soil solution equilibrium and alter the speciation and bioavailability of the metals. In the present context, therefore, the extraction and amendment of soil solution introduces changes that are unacceptable for the assessment of bioavailability.

Bioavailability of Metals in Terrestrial Ecosystems: Importance of Partitioning for Bioavailability to Invertebrates, Microbes, and Plants.
Herbert E. Allen, editor. ISBN 1-880611-46-5

Perhaps because of the difficulties outlined above, impacts on microorganisms have not been dealt with in some advanced risk assessment schemes and legislation, for example, the U.S. Environmental Protection Agency Part 503 rule (USEPA 1993).

This raises a question: Are soil microbes more sensitive to metals than other organisms such as higher plants and soil invertebrates? Few comparative studies exist, but there are indications that the answer to the question is "yes" (McGrath 1994). One example is the study of soil metal gradients at the Woburn Market Garden Experiment in the UK. Here, the effective population of a bacterium that fixes atmospheric N in symbiosis with clover plants, *Rhizobium leguminosarum* biovar *trifolii*, was extinguished and the whole soil microbial biomass was cut by half at soil metal concentrations that had little or no effect on crop plants or earthworms (Brookes and McGrath 1984; McGrath et al. 1988; Giller et al. 1989; McGrath 1994). Similarly, in 2 experiments at Braunschweig, Germany, the same rhizobial species became extinct in some high metal plots, but effects on crop growth and soil microbial biomass were not significant (Fließbach and Reber 1991; Chaudri et al. 1993). At Brunnby, Sweden, although no plant or invertebrate data were provided, there were small negative impacts on various microbial processes and populations at soil metal concentrations that were below the current European Union (EU) limits for soils that have received sewage sludge (Dahlin et al. 1997). These concentration limits are thought to protect against crop yield impacts and food chain impacts (Commission of the European Communities [CEC] 1986). In a review of published studies, a range of microbial assays and processes were shown to be more sensitive to Cu than a number of plant species and soil nematodes (Sauvé, Dumestre et al. 1998).

Another key issue is the deficiency and toxicity of essential elements such as Zn, Cu, and Mn. In soils, a sigmoidal cumulative frequency curve of soil metal concentrations is commonly observed (McGrath and Loveland 1992). This means that for each of these metals, there is a small population of soils in which biological activity may be limited by deficiency of essential metals (Figure 4-1). At the other extreme, there is another relatively small proportion of soils with elevated metal concentrations that are due to natural mineralization or anthropogenic activities. Most soils, however, show metal concentrations that are not limiting, and it can be presumed that most organisms have evolved to thrive across the range of "normal" concentrations characterized by the relatively flat part of each curve in Figure 4-1. Even in those soils with low contents of essential metals, organisms may have evolved mechanisms for solubilizing metals and tightly cycling them, so they rarely may be limited by trace-element deficiencies. The response of organisms to increasing metal bioavailability can be represented conceptually as shown in Figure 4-2. Although it is recognized that essential metals may be limiting in some circumstances, this review deals only with the toxicity portion of the dose–response curve because of the present context.

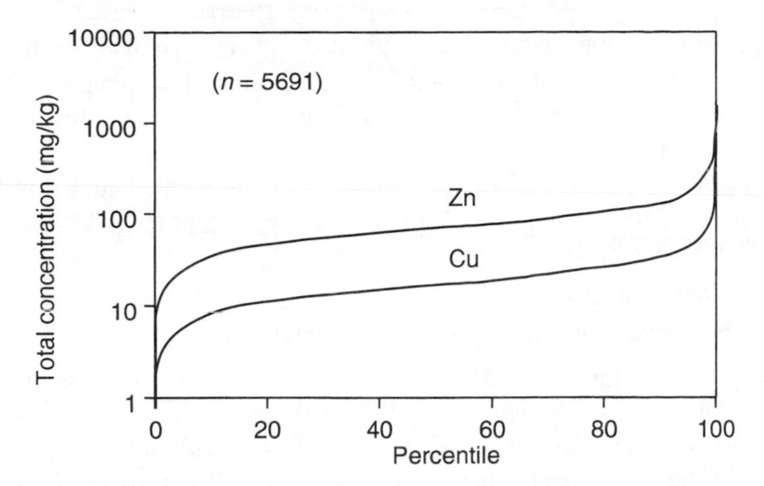

Figure 4-1 Percentile distributions of total Zn and Cu topsoil samples taken on a regular 5-km grid across England and Wales (data compiled from McGrath and Loveland 1992)

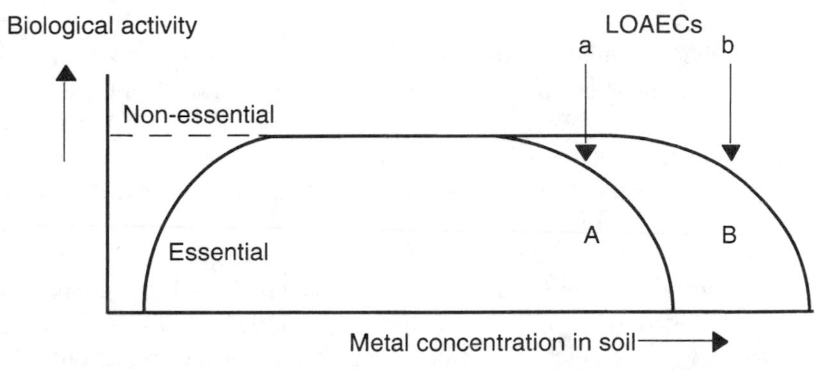

Figure 4-2 Conceptual diagram of the dose–response curve of 2 different metals in soil. One is non-essential and one essential; metals A and B have different buffering zones (plateau region where increasing metals have no effect) and different toxicity thresholds (lowest-observed-adverse-effect concentrations [LOAECs] a and b).

This chapter concentrates on metal bioavailability in aerobic terrestrial soils and does not include anaerobic systems, for which few data are available. Also, nematodes are not covered because some are not microscopic, and it is difficult to extract them specifically or to measure their activity in soil systems.

Data Availability

The availability of data on adverse effects of metals on soil microbes is limited for various reasons. Most of the available information relates to 5 types of measurement:

1) process measurements, such as respiration, N fixation, and nitrification;
2) total soil microbial biomass (assayed by methods such as fumigation-extraction, direct microscopic counts, or indirectly by adenosine triphosphate [ATP] or other measurements);
3) specific measurements of populations down to biovar level, for example, by estimating the most probable numbers in an extract of the soil (Vincent 1970; Colwell 1979);
4) soil enzyme activities; and
5) viable counts or enrichment culture measurements.

Table 4-1 illustrates for Cd how 3 major databases representing compilations of no-observed-adverse-effect concentrations (NOAECs) and lowest-observed-adverse-effect concentrations (LOAECs) vary in the type of microbial parameters chosen from world literature. Even though they reviewed about 30 years of published literature on ecotoxicological studies, there are relatively few entries for each of the toxic metals. Between 46 and 152 observations were included for Cd, but all of the values for soil contamination were reported only in terms of total metal concentrations in soil.

Furthermore, it is difficult to predict the solubility or bioavailability of total metal concentrations in the soils reported in these databases because the information on factors likely to affect solubility often is not present. The database of Witter (1992) contains information about the form of metal added, soil pH, percent clay, and percent organic C data, but these do not explain the variation in toxic threshold concentrations (Witter 1992). For example, the soils with the lowest pH, low clay content, and low percent C did not necessarily show the greatest toxicity. The database of Scott-Fordsmand and Pederson (1995) contains only the soil type and form of metal added to the soil, while Will and Suter (1995) published the soil type, soil pH, and percent organic C. Therefore, it is difficult to make predictions using these databases.

As can be seen from Table 4-1, there often are few observations in any one category. This also makes it difficult to construct relationships among soil metal content, soil properties, and a particular biological impact. It is not possible to combine them all because they consist of different organisms or microbial processes that themselves vary in sensitivity to metals (Bierkens et al. 1998). Results from the same type of assay in different laboratories may be hard to compare because of the lack of standardization of microbial test methods. It is also important to distinguish between the results reported from assays in which soluble metals salts are added to soil and measurements are made soon after and the results obtained in studies that monitor well equilibrated contaminated soils in the field. The bioavailability of the metals may be different in these 2 different situations, and so may the biological responses. The first involves fully soluble salt additions, which create a "disturbance" effect on the microorganisms; the second may consist of less soluble forms of metals, added over a long period, which can cause "stress" effects (see Giller et al.

Table 4-1 Soil microbial parameters included in the 3 ecotoxicology databases for Cd and the number of observations of each type[a]

Parameter	S-EPA[b]	DK-EPA[c]	ORNL[d]
Base respiration	20	—	—
C mineralization	7	13	4
Denitrification	9	—	4
Growth	25	—	—
Heterotrophic N_2 fixation	2	1	—
Microbial activity (unspecified)	—	1	—
Microbial composition or diversity	3	—	—
Microbial number	—	1	—
N mineralization	4	14	2
Nitrification	10	—	2
Size of population	23	1	—
Tolerance to metal	1	—	—
Soil enzyme activities:			
Acid phosphatase	4	—	2
Alkaline phosphatase	3	—	2
Amidase	3	1	—
Arylsulphatase	11	4	12
Asparaginase	—	1	—
Cellulase	—	1	1
Dehydrogenase	—	6	1
Hydrogen oxidation	3	—	—
Nitrate reductase	3	—	—
Phosphatase	—	1	7
Phosphodiesterase	2	—	—
Protease	1	—	—
Pyrophosphatase	3	1	—
Urease	15	5	9
Sub-total enzyme activities	*48*	*20*	*34*
Grand total	152	51	46

[a] Data from McGrath 1999.
[b] S-EPA = Swedish Environmental Protection Agency; Witter 1992.
[c] DK-EPA = Danish Environmental Protection Agency; Scott-Fordsmand and Pederson 1995.
[d] ORNL = Oak Ridge National Laboratory, USA; Will and Suter 1995.

1998 for discussion). Finally, there may be a lot of noise between laboratories because of the inherent difficulties associated with provision of any standard samples for microbial number or activity.

Another aspect that deserves attention is that, although many of the assays reported may have some relevance to the diversity and resilience of the soil as a functioning system, showing a change in diversity alone is not enough. Most reports to date fail to show links between diversity and essential soil functions. Biodiversity studies are becoming numerous, but until the relevance of the measurements to ecosystem function can be shown and predicted (Giller et al. 1998), it is not useful to assess the risks of metal contamination on the basis of simple changes in biodiversity. Also, there is some evidence that microbial diversity may actually increase when moderate metal stress is imposed (Giller et al. 1998).

Mode of Toxicity

It is now widely accepted that the toxic species in aquatic systems is the free metal ion (Sunda et al. 1978; Petersen 1982; Verweij et al. 1992; Campbell 1995; Allen and Hansen 1996; Lage et al. 1996), and it is assumed that the toxic species to soil microorganisms is also the free metal ion (Zevenhuizen et al. 1979; Bernhard et al. 1986; Hughes and Poole 1989; Morrison et al. 1989; McGrath 1994; Knight and McGrath 1995; Sauvé, Dumestre et al. 1998). It is likely that the bioavailable pool of metals includes the free metal ion in solution, plus a labile pool, which buffers the free metal concentration in solution. The labile pool may include both species on soil solids or soluble soil organic matter (SOM).

Many soil microbes may be exposed to the bulk soil solution and its free metal ion activity, but others may live in more specialized niches, in which the level of exposure is different. For example, some microbes live in colonies in biofilms that may complex metals and decrease their activity compared to the bulk solution. Others modify their surroundings as a result of excretion or other metabolic processes. For example, those that oxidize metal sulfides bring about the formation of acids, and these decrease the local pH and therefore increase metal solubility.

However, it is difficult to predict the metal activity in the numerous microsites in soil, so the Free Ion Activity Model (FIAM) in soil pore water is the prevailing theory to explain the response of microbes and soil processes to metals. This model assumes that heavy metal bioavailability is linked to the free ion activity in solution. This assumption is critical for this chapter and, as will be seen, relies on a few studies that cover a small number of metals to date.

Methods for Assessing Bioavailability of Metals

Given the above difficulties, 2 approaches have recently attempted to relate metal bioavailability to microbial effects on a more fundamental basis. One is to construct models of the soil factors that affect metal solubility and speciation in different soils, use literature data that report all the necessary parameters (where possible) to predict what the bioavailable concentration was, and relate this to the observed impacts. The other is to perform microbial impact measurements while at the same time measuring or computing metal speciation and bioavailability. This has brought about a new era in soil microbial toxicity assessment, as discussed in "Prediction approaches" below and "Linked speciation and ecotoxicity measurements" (p 78).

Prediction approaches

Active soil protozoa live in the soil pore water, and the mode of exposure to toxic metals is thought to be either through the water or by ingestion of contaminated bacteria. A toxicity test using *Colpoda inflata* was developed, and this protozoan was exposed to 1:4 soil–deionized water elutriates from mine waste–affected soils (Bowers et al. 1997). The free metal ion concentrations in the soil elutriates were calculated. However, in common with other assays employing protozoa, the medium contained dissolved organic materials (DOMs) that bind metals, so it altered the speciation in the samples. Also, the soils were contaminated with a number of metals, and the results could not be interpreted in terms of toxicity of any individual metal or metals.

Microbial respiration was measured by Lighthart et al. (1983) in 5 soils, with Cu or Cd added at 4 concentrations. The computer program GEOCHEM (Sposito and Mattigod 1980) was used to speciate the metals in a 1:1 soil–water filtered saturation extract of the soils. They did not fit the effect concentrations (ECs) to a dose–response curve, but the EC5 to EC10 was reported as $pCd^{2+} = 5$, and Cu gave inhibition when it reached $pCu^{2+} = 7$ to 8. However, it is unclear how the ECs were fitted in this early paper and whether the critical concentrations were the same across all soils. Other important soil factors such as Ca, organic matter, and nutrients that affect bioavailability and speciation of the metals were not considered in the equations.

Some of the more recent studies are retrospective. These select literature data on toxicity to microbial groups or processes, along with the reported soil properties that are important determinants of speciation. The free ion activity was then calculated from previously established regression relationships with soil parameters (Sauvé, McBride, Hendershot et al. 1995; McBride et al. 1997; Sauvé, McBride, Hendershot 1997, 1998b; Sauvé, McBride, Norvell, Hendershot 1997; Sauvé, Dumestre et al. 1998). This was necessary because almost none of the published biological studies have measured metal speciation in soil solution.

Sauvé, Dumestre et al. (1998) selected those studies that reported total metal and soil pH and in which the soils did not contain excessive concentrations of other toxic materials. This resulted in a small number of studies that included soil microbial measurements (8 papers). In particular, they noted that field-contaminated soils that fit these criteria were rare. They predicted pPb^{2+} and pCu^{2+} in soil solution from information given in the published papers. Prediction of effects on 5 of 7 microbial endpoints was improved by employing the estimated pPb^{2+} rather than total soil Pb. In the case of Cu, 10 of 13 assays gave better predictions using estimated pCu^{2+}. The suggested improvement can be illustrated by comparing the spread of percent inhibition of each assay to either pPb^{2+} or total soil Pb (Figures 4-3 and 4-4).

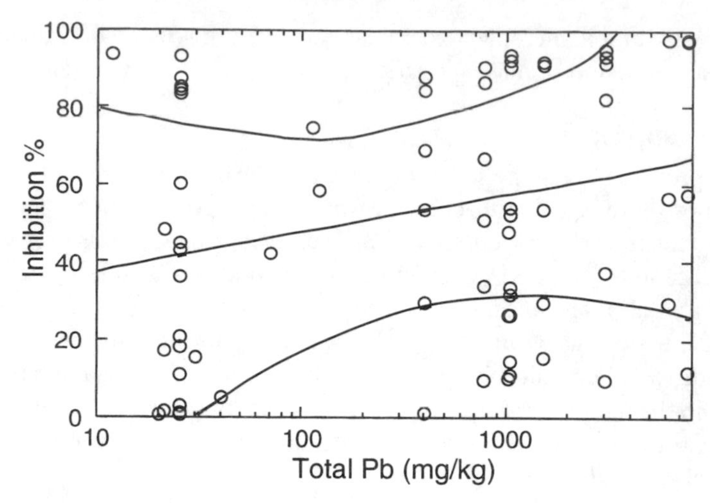

Figure 4-3 Inhibition of soil microbial assays in relation to published total soil Pb concentrations. The central line is a linear regression ($R^2 = 0.08$, $n = 75$), and the curves are 90% confidence intervals. Assays as in Figure 4-4 (redrawn from Sauvé, Dumestre et al. 1998).

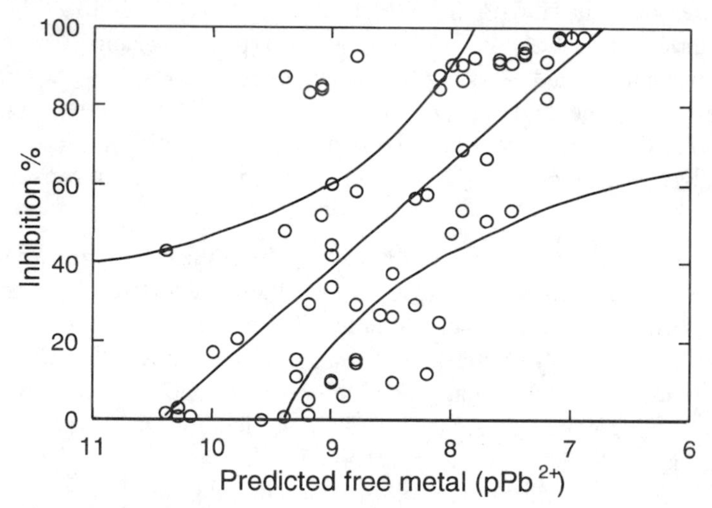

Figure 4-4 Inhibition of soil microbial assays in relation to calculated pPb^{2+} in soil solution. The central line is a linear regression ($R^2 = 0.48$, $n = 67$), and the curves are 90% confidence intervals. Assays were from literature studies of soil respiration, nitrification, N mineralization, urease activity, dehydrogenase activity, and mycorrhizal growth (redrawn from Sauvé, Dumestre et al. 1998).

The pPb^{2+} activity in solution predicted to give 50% inhibition was 8.1 ± 1.0 (equivalent to 1.6 mg Pb^{2+}/L), and for pCu^{2+} this was 8.2 ± 1.3 (0.4 mg Cu^{2+}/L).

Despite the improved prediction, a number of difficulties are associated with this approach. Uppermost is the wide data spread, which was worse for Cu^{2+} than for Pb^{2+} (Figures 4-4 and 4-5). In addition, the spread in the calculated Pb^{2+} is larger

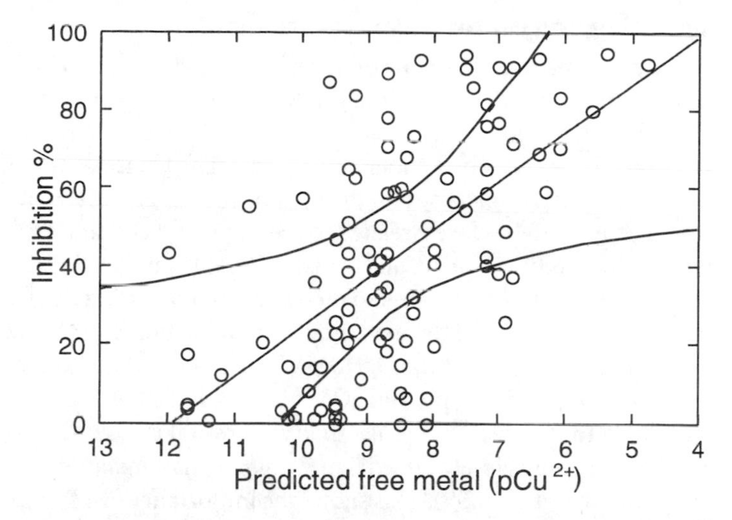

Figure 4-5 Inhibition of soil microbial assays in relation to calculated pCu^{2+} soil-solution concentrations. The central line is a linear regression ($R^2 = 0.36$, $n = 103$), and the curves are 90% confidence intervals. Assays were from literature studies of microbial biomass, soil respiration, nitrification, denitrification, C mineralization, N mineralization, urease activity, ethylene production, rhizobia survival, and Biolog tests (redrawn from Sauvé, Dumestre et al. 1998).

than the soil total Pb (Figures 4-3 and 4-4). Note also that the errors on the EC50 values are 1 log unit or more, which means that the EC50 varies by 2 orders of magnitude. At the fitted pCu_{50}^{2+}, the observed percent inhibition for the microbial assays ranged from 0 to nearly 100%. At pCu^{2+} of 7 or lower (i.e., higher metal concentrations), nearly all assays were more than 40% inhibited. Similarly, at above pCu^{2+} of 10 (low metal concentrations), most assays were less than 20% inhibited (Figure 4-5). The variation in these assessments could be for 4 main reasons:

1) real differences in sensitivity to metals between different groups, species, or between different microbially mediated processes, which seems to be greater for Cu than for Pb;
2) errors associated with the fact that metal speciation was not actually measured in the experiments linked to the soil assays;
3) failure to include other factors that may have affected bioavailability and speciation in particular soils; and
4) interlaboratory variation in the microbial assay methods.

The latter factor probably was not too important in this study because there were only a few studies that measured the same microbial assays. However, the influences of the other factors are so great, and the spread in data so large, that this approach of using literature data from disparate laboratory studies may not be useful for regulatory application.

Linked speciation and ecotoxicity measurements

Free Cu^{2+} activity was measured by cupric ion-selective electrode (ISE) in field soils contaminated specifically with Cu and equilibrated for up to 50 years or more (Dumestre et al. 1999). Various media were used to measure mineralization of added glucose and viable microbial population counts. The length of the lag period (LP) between the addition of glucose and the start of the exponential phase of mineralization is known to be closely related to the soil metal contamination (Bewley and Stotzky 1983; Doelman and Haanstra 1984; Dahlin et al. 1997; Nordgren et al. 1998). The LP was found to increase with increasing Cu (Dumestre et al. 1999), and they also measured the maximum mineralization rate (MMR, slope of the exponential phase of mineralization). However, the MMR was affected by soil factors other than Cu, specifically pH and SOM content. Both MMR and LP were more closely related to the measured pCu^{2+} than either total dissolved Cu or total Cu in soil, but MMR was more closely linked to the organic matter status than any of the Cu pools (Dumestre et al. 1999). This shows the importance of LP as an indicator of Cu toxicity that is relatively unaffected by other soil properties and the poor utility of MMR as an indicator of metal bioavailability because of its sensitivity to the total pool of substrate present (SOM). The LP was 0 or close to 0 in soils with low Cu concentrations ($pCu^{2+} > 10$, <0.0064 mg Cu^{2+}/L) and increased to 3.5 hours at pCu^{2+} of 8 (0.64 mg Cu^{2+}/L) and to 7 hours at pCu^{2+} of 6 (64 mg Cu^{2+}/L) (Figure 4-6). Increased LP is a measure of the inhibition of mineralization, and if an 8-hour LP is taken as the maximum inhibition, 50% inhibition would be at pCu^{2+} of about 8 for these soils. However, this may not be a good parameter to do predictions with because a different set of soils may give a different absolute range of LPs, making it difficult to calculate a 25% or 50% inhibition point that is valid for all soils.

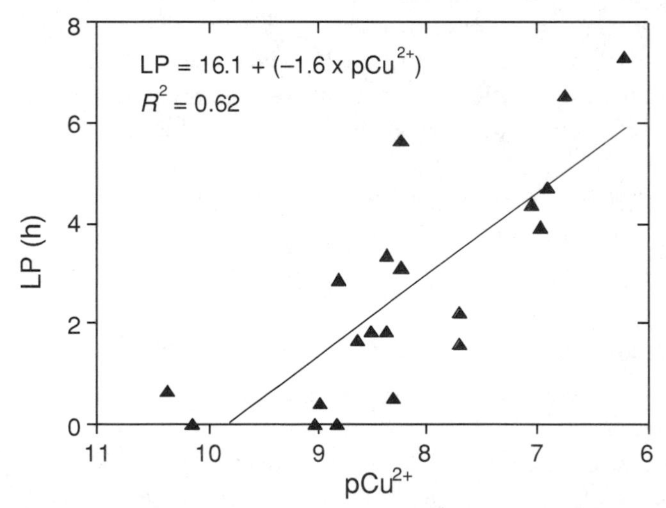

Figure 4-6 Lag period for glucose mineralization, in relation to soil pCu^{2+} based on soils from field sites contaminated with Cu from New York and Denmark (redrawn from Dumestre et al. 1999)

In the same study, the 7 soils from New York were plated out on specific media for counting viable colony-forming units of bacteria, actinomycetes, and fungi. Plates with added Cu were used to estimate the Cu tolerance of these groups also. However, because of the well-known problems of metals, especially Cu, being bound by organic media components (Angle et al. 1992) and the inherent selectivity of plate counting methods (Amann et al. 1995), the results were somewhat equivocal. Only one of these soils had high pCu^{2+}, and more Cu-tolerant organisms from all groups were isolated from this soil. The bacteria, which grew on the plates, were more sensitive to Cu than the actinomycetes and fungi, and the number of fungi isolated was slightly smaller in the most contaminated of the soils.

The nitrification potential of soils contaminated with either Cu or Pb from fields in New York State, Québec, and Denmark was measured by oxidation of added ammonium (Sauvé et al. 1999). Free Cu^{2+} activity was measured by cupric ISE and chemical speciation of Pb^{2+} by differential pulse-anodic stripping voltammetry (DP-ASV) in the field soils that had been contaminated for 10 years or more. However, pCu^{2+} did not affect nitrification in this study, and it was concluded that this process is sensitive to many environmental parameters (particularly pH and SOM), and therefore is not a good indicator of soil metal contamination. It is important to realize that this also may be the case with a number of potential bioindicators because they may be

1) insensitive to metals,
2) intolerant to a wide range of soil pH, and
3) sensitive to other soil factors such as soil pH and organic matter content.

A different approach was taken by McGrath et al. (1999). They extracted soil solution with inert hollow fiber probes (Knight et al. 1998) inserted in soils from 2 separate metal gradients from 2 field sites in Germany. One site had been in arable production for most of this century (termed "old arable"), and the other was cleared from forest around 1950 ("ex-woodland"). The soils had been treated with sewage sludge for 10 years until 1990, and the soil metal concentrations, apart from Zn, were below the EC limits (CEC 1986). An ion exchange equilibrium (IEE) technique was used to measure the free Zn^{2+} concentration in the soil solutions (Holm et al. 1995). The biological impact of the Zn^{2+} was evaluated by exposing a common soil bacterium, *Pseudomonas fluorescens* strain 10586s pUCD607, with the *lux* insertion on a plasmid (Paton et al. 1997) to the soil solution samples and measuring the decrease in bioluminescence compared with soil solutions from uncontaminated plots at each site. The presence of a toxic metal at concentrations that impose a deleterious impact on cell metabolism causes a decline in the output of bioluminescence of *lux*-marked bacteria.

When the results, expressed as percentages of the luminescence of control soil solutions, were related to the total Zn concentrations in the soils, different relationships were found for each soil (Figure 4-7). However, when expressed against the pZn^{2+} in soil solution, one curve fits both soils and explained 73% of the variance in

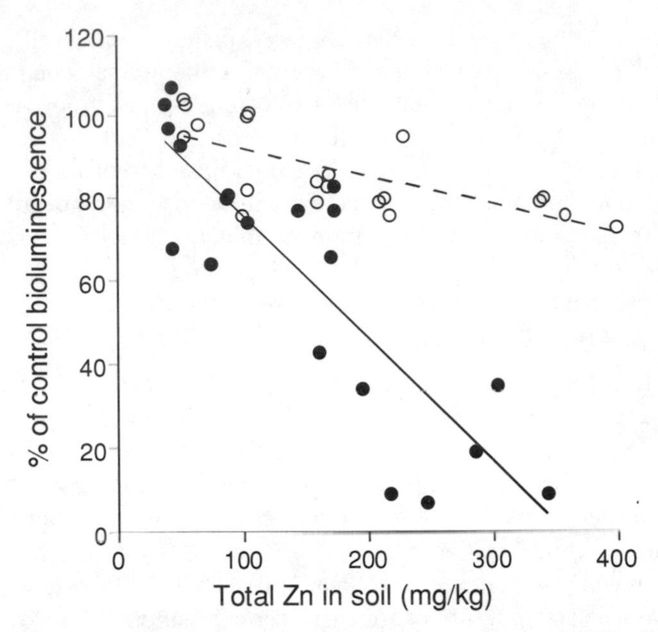

Figure 4-7 Relationships between the bioluminescence of *lux*-marked *Pseudomonas fluorescens* 10586s pUCD607 and total concentrations of Zn in soils from a gradient of Zn contamination in 2 soils from Braunschweig, Germany (● ex-woodland soils, ❍ old arable soils). Plots with no sludge applied were used as controls. (Reprinted with permission from McGrath SP, Knight B, Killham K, Preston S, Paton GI. 1999. Assessment of the toxicity of metals in soils amended with sewage sludge using a chemical speciation technique and a *lux*-based biosensor. *Environ Toxicol Chem* 18:659–663. Copyright SETAC, Pensacola, FL, USA.)

the data set (Figure 4-8). This curve constitutes the toxicity section of the idealized dose–response curve that applies to both sites (Figure 4-2) with an EC25 value of $pZn^{2+} = 4.53 \pm 0.13$ (equivalent to 1.9 ± 0.6 mg Zn^{2+}/L). The EC50 value was 4.03 $pZn^{2+} \pm 0.08$ (6.1 ± 1.2 mg Zn^{2+}/L). The effect is unlikely to be due to the slightly lower pH of the ex-woodland soil because the bioluminescence is optimal at about pH 5.5 and is stable across a wide range of solution pH (Paton et al. 1997), decreasing markedly due to acidity only below about pH 4 (Parker et al. 1995).

Soil-solution free metal ion activity may be a useful parameter to investigate the response of soil organisms to toxic metals (McGrath 1994; Sauvé et al. 1999; Dumestre 1999). This measured species in soil solution is the result of those soil properties that determine bioavailability of metals (e.g., pH, SOM, and DOM). In this case, without the need to predict either the partitioning of Zn between soil solid and solution or the speciation of Zn in solution, pZn^{2+} may be a step toward a unified method for assessing Zn availability in soil solution to microbes across different soil types.

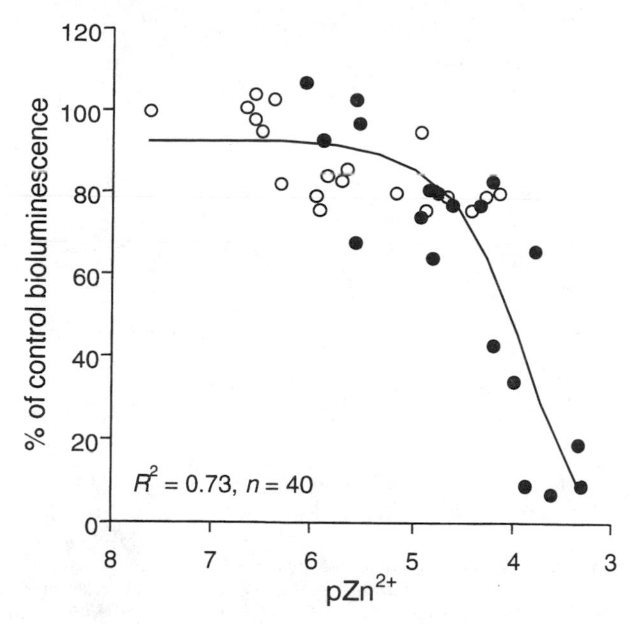

Figure 4-8 Relationships between the bioluminescence of *lux*-marked *P. fluorescens* 10586s pUCD607 and the pZn^{2+} in soil solutions extracted from a gradient of Zn contamination in 2 soils from Braunschweig, Germany (● ex-woodland soils, ❍ old arable soils). $R^2 = 0.73$ for both pZn^{2+} and soluble Zn concentrations in soil solutions, $n = 40$. Plots with no sludge applied were used as controls. (Reprinted with permission from McGrath SP, Knight B, Killham K, Preston S, Paton GI. 1999. Assessment of the toxicity of metals in soils amended with sewage sludge using a chemical speciation technique and a *lux*-based biosensor. *Environ Toxicol Chem* 18:659–663. Copyright SETAC, Pensacola, FL, USA.)

This method was used on another gradient sampled at the Gleadthorpe experiment in the UK, which had received metal-contaminated sewage sludges 10 years or more before sampling (Chaudri et al. 1999). The sludges were contaminated with Zn, Cu, or a mixture of Zn and Cu (Zn+Cu). Even the so-called "Cu" sludge contained Zn in relatively large concentrations. A challenge in this study that is common to most really polluted sites was the interpretation of the impact of the mixed metal pollution present. Because much of the Cu in soil solution was complexed with dissolved organic C (DOC), it was not possible to estimate small concentrations of free Cu^{2+} by IEE. However, the toxic threshold for Cu was not reached (see the next 2 paragraphs), so speciation analysis of Cu was unnecessary.

The organisms used in the luminescence-based bioassays were *Escherichia coli* HB101 pUCD607 and *P. fluorescens* 10586rs pUCD6 (Chaudri et al. 1999). It was clear that the observed effects were in fact due to Zn, even in the plots with Cu or Zn+Cu treatments (Figure 4-9). The effect of the Cu in soil solution was shown to be insignificant in plots that received Cu or Zn+Cu-contaminated sludges by both multiple linear regression and stepwise multiple linear regression analysis. Even the

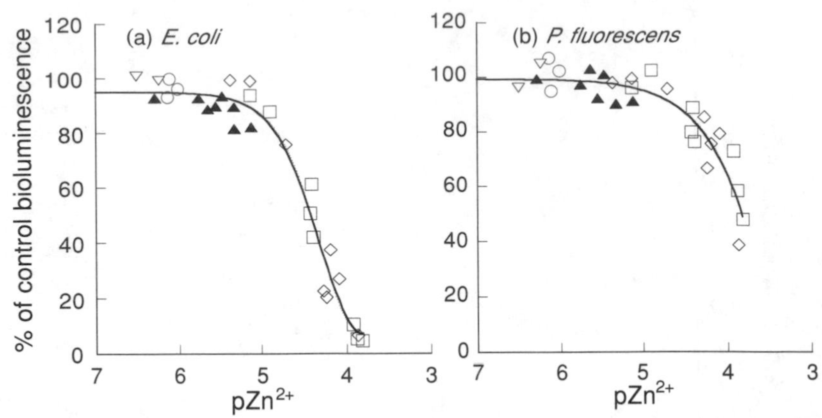

Figure 4-9 Relationships between bioluminescence of (a) *E. coli* HB101 pUCD607 and (b) *P. fluorescens* 10586rs pUCD6 and pZn^{2+} in soils treated with Zn-, Cu-, or Zn+Cu-contaminated sewage sludges more than 10 years previously. R^2 *E. coli* = 0.96, n = 30, R^2 *P. fluorescens* = 0.88, n = 30. Note: solid symbols represent Cu treatments. Treatments: ❍ No sludge control; ▽ Uncontaminated control sludge; ☐ Zn-contaminated sludge; ◊ Zn+Cu-contaminated sludges; ▲ Cu-contaminated sludge. (Reprinted with permission from Chaudri AM, Knight BP, Barbosa-Jefferson VL, Preston S, Paton GI, Killham K, Coad N, Nicholson FA, Chambers BJ, McGrath SP. 1999. Determination of acute Zn toxicity in pore water from soils previously treated with sewage sludge using bioluminescence assays. *Environ Sci Technol* 33:1880–1885. Copyright American Chemical Society.)

soil samples with the largest concentration of total Cu in solution (620 µg/L) gave percentage decreases in bioluminescence of only 18% and 9% for *E. coli* and *P. fluorescens*, respectively.

The response of the 2 luminescence-based bioassays to soil pore water from plots that received a mixture of Zn+Cu-contaminated sludges were similar to that from plots receiving only Zn-contaminated sludges with similar Zn^{2+} concentrations. This suggests that the decrease was due to Zn^{2+} in solution and not to Cu (Figure 4-9). EC25 and EC50 values were pZn^{2+} of 4.71 ± 0.058 and 4.41 ± 0.027 for *E. coli* and 4.17 ± 0.045 and 3.84 ± 0.037 for *P. fluorescens*, respectively. R^2 for pZn^{2+} was 0.96 and 0.88 for *E. coli* and *P. fluorescens*, respectively. Correlations with soil total Zn in the same experiment were still strong (R^2 = 0.84 and 0.75, respectively). However, this is to be expected when samples are from a site with one soil type. A more stringent test would be studies on soils from different soil types.

Incorporation of Bioavailability into a Regulatory Framework

Metal toxicity to soil organisms should be evaluated in the same way as has been done for aquatic systems and ideally should be based on results from tests on representatives of each of the groups of organisms listed previously. In the case of soil microbes, the differences in sensitivity of different groups and microbial processes may contribute to the large variability encountered in studies that include a number of diverse types of measurement (e.g., Figure 4-3). If we express the impacts of metals against a measure of bioavailability, there is an improvement, but there still is variation due to biodiversity and errors in the prediction of the soluble species of metals present (Figures 4-4 and 4-5).

On the other hand, assays performed with one or few organisms, such as bacteria, show less variability, especially when the metal species are measured at the same time the experiments are performed (Chaudri et al. 1999; McGrath et al. 1999). In these examples, the biosensors are being used as indicators of biological response, specifically toxic effects of metal in soil solution on cell metabolism (Paton et al. 1997). Whether they represent all the organisms present in soil is unlikely, given the potential biological diversity.

As indicated above, there is a lack of data on all the groups and processes that can be related to bioavailable concentration of metals and measurements that can be statistically evaluated to yield consistent metal threshold values for each. In any case, it is unlikely that all groups and processes will be tested for all metals in the near future.

How do the data that are presently available compare with each other? There is a degree of agreement in the EC values derived from the studies discussed above. Obviously, there is more information for Cu and Zn than for other metals. It appears that pZn^{2+} values of 4.47 to 4.09 (2.22 to 5.31 mg Zn^{2+}/L) are associated with a 25% to 50% reduction in activity of various microbes and microbial processes (Table 4-2). In the case of Cu, the data fall into 2 groups:

1) those of Sauvé, Dumestre et al. (1998) and Dumestre et al. (1999), with EC50 pCu^{2+} values around 8; and
2) recent bacterial biosensor work (Vulkan et al. 2000), with EC50 for pCu^{2+} of 2 different bacterial biosensors of around 5.

In the first group, Sauvé, Dumestre et al. estimated the pCu^{2+} affecting various microbial endpoints, while Dumestre et al. measured the effect of pCu^{2+} on soil respiration using an ISE. The bacterial biosensors were employed in long-term Cu-polluted soils, and pCu^{2+} was measured by ISE. It is possible that the differences in sensitivity are due to a combination of the following:

- The Cu studies of Sauvé, Dumestre et al. (1998) and Dumestre et al. (1999) mainly used whole soil assays and had many species present, while those of Vulkan et al. (2000) involved 2 bacterial species that may be less sensitive to Cu.
- Different calibration and computation methods were used to measure pCu^{2+}.
- Exposures during the tests ranged from short to long term (Vulkan et al. 2000).

Table 4-2 Summary of available critical concentrations of free metal ions in soil solution in relation to microbial effects

					pM^{2+}		M^{2+} mg/L		
Metal	Microbial assay	Nr. observations	pM^{2+} undefined EC	mg/L undefined EC	EC25	EC50	EC25[a]	EC50[b]	Source
Cu^{2+}	Respiration	15	7 to 8	0.006 to 0.0006	—	—	—	—	Lighthart et al. 1983
Cu^{2+}	Various	103	—	—	—	8.2	—	0.0004	Sauvé, Dumestre et al. 1998
Cu^{2+}	Lag period (LP)	20	—	—	—	8.0[b]	—	0.00064	Dumestre et al. 1999
Cu^{2+}	*E. coli*	22	—	—	5.69	4.78	0.13	1.05	Vulkan et al. 2000
Cu^{2+}	*P. fluorescens*	22	—	—	5.88	5.19	0.08	0.41	Vulkan et al. 2000
Cu^{2+} *mean*		—	—	—	—	*6.30*	—	*0.04*	
Zn^{2+}	*P. fluorescens*	40	—	—	4.53	4.03	1.93	6.10	McGrath et al. 1999
Zn^{2+}	*P. fluorescens*	30	—	—	4.17	3.84	4.42	9.45	Chaudri et al. 1999
Zn^{2+}	*E. coli*	30	—	—	4.71	4.41	1.27	2.54	Chaudri et al. 1999
Zn^{2+} *mean*		—	—	—	*4.47*	*4.09*	*2.22*	*5.31*	
Pb^{2+}	Various	67	—	—	—	8.1	—	0.0016	Sauvé, Dumestre et al. 1998
Cd^{2+}	Respiration	15	5	1.12	—	—	—	—	Lighthart et al. 1983

[a] Calculated from pM values.
[b] EC50 not absolute; this represents half of the range of the observed LPs.

For Cd, 5% to 10% inhibition occurred at pCd^{2+} of 5 (1.12 mg Cd^{2+}/L), and EC50 for pPb^{2+} was about 8 (0.0016 mg Pb^{2+}/L).

In the future, application of these critical values will depend on the following developments:

1) Gathering new information on critical values for other potentially toxic metals on the basis of pM^{2+} in soil solution
2) Obtaining data to show whether the critical values obtained hold for a wide range of soils. To date, the biosensor studies have been accomplished on gradients of Zn on 3 soils (Chaudri et al. 1999; McGrath et al. 1999), while the Cu study of Vulkan et al. (2000) included different sources of contamination in soils from 14 locations in different countries (Table 4-2).
3) Determining whether organisms exposed over longer periods to the soil solid phase react in the same way as those exposed only to soil solutions
4) Quantifying the effects of interactions between metals.

It is the domain of regulators to decide whether the NOAECs, LOAECs, EC25, or EC50 concentrations would be regarded as the control points in future legislation. There are technical problems in establishing NOAECs (Chapman et al. 1996), and LOAEC values depend on the exact concentrations used in tests (Witter 1992). In theory, EC values are straightforward to produce, but legislative acceptance will be enhanced if the biological endpoints can be shown to reflect the impacts of chronic metal pollution rather than simple acute toxicity. Generally, studies of effects of metals on indigenous microbial populations often report short-term acute effects rather than long-term chronic effects. Chronic effects take many years, if not decades (Bååth 1989; Chaudri et al. 1992, 1993) to occur, even in highly polluted soils, making early detection of problematic soils difficult. If microbial biosensors can be shown to be proxy indicators of chronic effects, then they will be extremely useful tools in the identification of potentially toxic soils, providing an early warning system. This information, once collected, will be valuable in systems that use basic ecotoxicological dose–response data in a uniform way (e.g., based on free metal species in soil solution).

It is known that loss of microbial diversity can occur at concentrations of bioavailable metals that are lower than those associated with the selection of metal tolerance in some groups (Giller et al. 1998). Tolerance has been suggested as a useful indicator of microbial response to metals (Díaz-Raviña et al. 1994; Díaz-Raviña and Bååth 1996a, 1996b); however, it appears that metal tolerance may become common only as a result of extreme selection, that is, at concentrations above those where community change and loss of function has occurred. Biological function is impaired at the high concentrations of bioavailable metals associated with former smelter sites, as evidenced by the buildup of undercomposed organic matter in seriously impacted soils (Tyler 1983).

Rutgers et al. (1998) and Rutgers and Breure (1999) used soil extracts from Zn-contaminated soils inoculated onto wells in Biolog plates along with metal treatments to assess pollution-induced community tolerance. The method measures the utilization of different sole C sources by the microbes in the extract. However, it is worth noting that only culturable organisms, mainly bacteria, will grow on the plate (<1% of the microbial community). Also, selection may take place on the plate, and the observed resistance may be represented by isolates that are rare in the soil environment. Rutgers et al. (1998) showed results for 2 substrates that have a significantly increased EC50, and it can be estimated that these take place on previous exposure to about 250 mg/kg Zn or more in the soil. Results of Rutgers and Breure (1999) indicate that there is a shift in the general median EC50 value from about 250 to 2500 mg/kg Zn in soils with increasing Zn added in the field. These lower EC50 values are of interest because they are less than those of Díaz-Raviña et al. (1994) and Díaz-Raviña et al. (1996a, 1996b). Bååth et al. (1998) also found increases in bacterial community tolerance in soil extracts from soils containing more than 200 mg/kg Zn, using short-term thymidine incorporation tests after adding metal salts to soil extracts.

As noted above, more studies are needed to establish critical values based on soluble metal concentrations in soil solutions from soils with different physical and chemical characteristics. It remains to be shown whether this approach yields more uniform critical values across a large number of different soils than do soil total concentrations. An important improvement in risk assessment could be offered by moving from legislation based simply on total concentration in soil to free metal ion concentrations in soil solution or free metal ion concentrations plus an as-yet-undefined "labile" pool which backs up the solution concentration.

Systems based on free metal ion concentrations could be used in practice in 2 ways. First, the effects could be determined directly, for example, biosensors could be applied to soil solutions from a contaminated site (Paton et al. 1997; Knight et al. 1998; Chaudri et al. 1999; McGrath et al. 1999), and the free ion concentrations could be determined in the same samples of solution. In the case of multi-metal contamination, the results for each metal should be compared with the preestablished critical values for each element.

Second, it is more likely that the measurement phase discussed above will act as a kind of calibration. Once the biologically critical pM^{2+} values in solution are established, it will be possible to either measure or predict chemically whether the critical concentration of the toxic species has been reached in a given situation. Measurement of free metal species is difficult and not amenable at present to employment in routine laboratories. This makes it more likely that prediction will be used.

Again, there are 2 possibilities. One is that after measuring all the concentrations of metals and anions, DOC, and pH, the concentrations of free metal species can be predicted using models such as WHAM (Tipping 1994) or GEOCHEM-PC (Parker et al. 1995). Alternatively, the easiest method would be to use algorithms that relate important soil properties such as total metal concentration, pH, SOM, oxide, and clay contents to the likely free ion concentration in soil solution (e.g., the methods suggested in this book). However, whether variations in the mineral forms of metals present in different soils give rise to different relationships because of important differences in their solubility still needs to be established. Finally, the issue of "aging" of metals added to soil will also have to be dealt with by a modeling approach. In the absence of such a model, we presently can deal only with soils that have been polluted for a long time, that is, those that presumably have reached an equilibrium.

Multidisciplinary teams can make progress in these tasks by having the skills to perform appropriate microbial measurements alongside measurement of solubility and speciation and/or chemical computer modeling. In future, these approaches could be very powerful for regulating metal loads in soils on the basis of the toxic bioavailable concentrations.

Chapter 5

Bioavailability of Metals to Soil Invertebrates

Willie J.G.M. Peijnenburg
National Institute of Public Health and the Environment

Invertebrates are among the major components of soil biomass and play an important role in maintaining the structure and fertility of soil. Invertebrate-mediated processes such as drainage, aeration, and incorporation and degradation of organic matter are important in improving soil quality (Edwards and Lofty 1977; Barber et al. 1998). Moreover, invertebrates are an important part of the terrestrial food web and can constitute a significant component of the diet of birds, small mammals, reptiles, and other soil-inhabiting biota. Because of these characteristics, invertebrates have been adopted as important indicator organisms for assessing potential impacts of chemicals to soil organisms and to organisms in the terrestrial foodweb. More specifically, standardized tests have been developed to quantify effects of chemicals on earthworms (Organisation for Economic Cooperation and Development [OECD] 1984). Recently, acute, chronic (reproduction), and field toxicity tests on earthworms have been standardized by the International Organization for Standardization (ISO 1998). In ecotoxicology tests, effects on biota usually are related to the total metal content in the soil. However, the relationship between the total metal content and its positive or negative effect on soil biota is not straightforward. In natural systems, the exposure to metals depends not only on the total metal content but also on the composition of the system, on a number of environmental conditions (e.g., pH, temperature, concentration of competing ions such as Ca, and concentration of complexing ligands in solution), and on the chemical form of the element (with kinetically "stable" metal species being less toxic in general than species that are in true equilibrium with free metal ions). In addition, characteristics of the organism itself might play an important role.

Because of the standardization of test methods, little attention has been paid to aspects of substrate quality and to duration of exposure. First, insufficient attention is paid to the rate and extent at which tissue concentrations respond to substrate concentrations. There is little consideration of the factors that modulate the availability of metals, especially in the case of soils. Therefore, it is critical to assess metal bioavailability using soils in which the metal contaminant has reached a certain chemical steady state with respect to soil processes such as chemisorption, mineral equilibria, and organic matter complexation. Many instances have shown that metals from freshly salt-spiked soils are much more toxic than equivalent metal

Bioavailability of Metals in Terrestrial Ecosystems: Importance of Partitioning for Bioavailability to Invertebrates, Microbes, and Plants.
Herbert E. Allen, editor. ISBN 1-880611-46-5

concentrations in field-collected soils. A recent study of *Eisenia fetida* showed, for example, that the metal burden of field-collected soils is less available than the same concentrations spiked into artificial soil media (Spurgeon and Hopkin 1995).

Much of the variability between soil bioassays arises from the drastic influence of soil properties such as pH, organic matter, Ca status, or cation exchange capacity (CEC) (Kennette et al. 2001). Unfortunately, some of the effects and trends are metal or species specific, that is, lower pH increases earthworm bioaccumulation of Cd and Zn (Beyer et al. 1982; Ma 1982; Ma et al. 1983; Beyer et al. 1987; Perämäki et al. 1992), and it also is common knowledge that the toxicity of metal ions for plants growing in soils decreases with increasing pH (Lexmond 1980), whereas the toxicity of metal ions often decreases with increasing pH for plants grown in a nutrient solution (Lexmond and Van der Vorm 1981). In contrast, Cu uptake by earthworms does not appear to be influenced by soil pH (Ma 1982; Ma et al. 1983). Soil Pb bioavailability to earthworms is more variable; some studies show a strong increase in Pb bioavailability under acidic conditions (Ma et al. 1983; Ma 1987; Morgan and Morgan 1988), while other studies show no effect of soil pH (Andersen 1979; Helmke et al. 1979; Beyer et al. 1982) or some slight increases in bioaccumulation of Pb under low pH conditions (Perämäki et al. 1992).

Uptake of Pb by earthworms is also believed to have an unusual relationship with soil Ca. Andersen (1979) found that increased tissue uptake of Ca is followed by elevated bioaccumulation of Pb and noted that species with higher Ca turnover rates (such as *Lumbricus terrestris*) appeared to bioaccumulate more Pb than species with lower Ca turnover rates (such as *Allobophora*). It has also been suggested that soil Ca competitively suppresses Pb uptake in earthworms and that Pb bioaccumulation by earthworms is substantially lower in specimens sampled in calcareous soils (Andersen 1979; Morris and Morgan 1986). Soil organic matter (SOM) was also shown to be associated with earthworm bioaccumulation of Cu (Ma 1982), Pb (Ma et al. 1983), and Cd (Morgan and Morgan 1988). Furthermore, most soil properties are autocorrelated to various degrees (Beyer et al. 1987; Morgan and Morgan 1988; Basta et al. 1993), and it becomes difficult to isolate the actual source of an effect from the interdependent factors. Because it is difficult to account for all the possible factors influencing the relative bioavailability of the soil total metal burden, metal bioavailability is often approximated using chemical extraction techniques that are intended to integrate the effects of the various potentially interacting soil properties. The sequential extractions try to find a procedure and reagent that will extract a fraction of the soil metal pool, which can later be correlated to biological uptake, toxicity, and potential risk. Unfortunately, there is a plethora of proposed reagents that usually have not given any consistent results. The proposed extraction procedures vary among test species, metals, and soil systems. Furthermore, debate persists as to whether chemical measurements can provide valid information on the actual availability of the metals to soil organisms (Morgan and Morgan 1988).

Second, with regard to the aspect of time, invertebrate testing is proposed as an alternative bioassay where a soil-dwelling organism is used to determine the potential biological availability of a given soil metal burden. Such a bioassay would reflect a soil exposure that integrates the effects of the chemical properties of a given soil over a determined period. Ecotoxicological testing of soils with earthworms is typically carried out by recording health effects after a relatively short exposure time (e.g., 3 weeks). However, as indicated by Hopkin (1989) and Sheppard et al. (1997) for elements such as Cd, there is good evidence that the kinetics are slow and that accumulation of the element may continue for the life span of the organism (more than several months). Steady state would be the preferred condition for interpreting test results and for allowing extrapolation of laboratory toxicity data to field conditions. Many toxicity data, however, are not obtained at or near steady state, and steady-state tissue concentrations can be much higher than those in a short-term bioassay.

Metal bioavailability is both metal and species dependent, and it also is dependent on the interaction between a chemical and an organism as a function of time (Rand 1995). In view of the time issue, bioavailability should preferably be addressed as a dynamic process. The dynamic approach of "bioavailability" in soil should comprise at least 2 distinct phases: a physicochemically driven desorption process (the chemical supply), and a physiologically driven uptake process (McCarthy and Mackay 1993). The latter requires identification of specific biotic species as an endpoint. Hamelink et al. (1994) referred to these processes as "environmental availability" and "environmental bioavailability," respectively. Because the level in the organism must reach some threshold value at the target site (e.g., cell membranes for narcotic agents) before effects start to occur, "toxicological bioavailability" was defined by Hamelink et al. (1994) as being the final determinant of toxicity. Lanno et al. (1998) applied the concept of body residues as a tool for assessing toxicological bioavailability, taking into account that, once taken up, metals may be partitioned into either biologically available, biologically unavailable, or storage fractions. Biologically available metals, in their turn, can participate in essential metabolic functions or, in the case of nonessential elements or excess essential metals, can contribute to toxicity. The use of body residues for the appropriate species reduces uncertainties in risk assessment procedures (Van Wensem et al. 1994; Van Straalen 1996). However, even when there is uptake, there is not necessarily toxicity because the organisms may sequester metals and thus avoid physiological impact. In addition, organisms may change their environment, thereby altering the metal distribution in soil (e.g., plant-root exudates) and inside the organism. An example of this may be found for the collembola *Orchesella cincta.* Van Straalen and van Meerendonk (1987) conclude that, given the large amounts of Pb processed by *O. cincta* populations, Collembola may have a long-term solubilizing effect on Pb in the soil. Therefore in some species, external metal concentrations, rather than internal (noncomplexed) concentrations, may directly determine toxic effects.

Environmental conditions and organism-specific uptake routes play crucial roles in the whole context of bioavailability because they determine the steady-state status. The environmental conditions, moreover, play a role in the survival and well-being of soil organisms. Soil organisms potentially have different uptake routes. There is evidence for predominant porewater uptake of organic substances by soft-bodied animals (Belfroid et al. 1996), but because of their complex physicochemical behavior, such evidence at present is only circumstantial for metals (Spurgeon and Hopkin 1996). Accumulation data in Lumbricidae are commonly used in risk assessments for invertebrate-eating animals (Romein et al. 1991; Balk et al. 1993). This approach neglects the possible differences in bioaccumulation between invertebrate species. Variability can be expected because there is much variation in morphology, physiology, behavior, habitat, and food preference among invertebrate species. For a proper risk assessment, knowledge about accumulation patterns of other invertebrate species is needed to obtain more insight into the representativeness of the use of Lumbricidae bioaccumulation data for assessing both risk and bioavailability differences among species. The 3 principal processes involved in the concept of bioavailability are depicted in Figure 5-1.

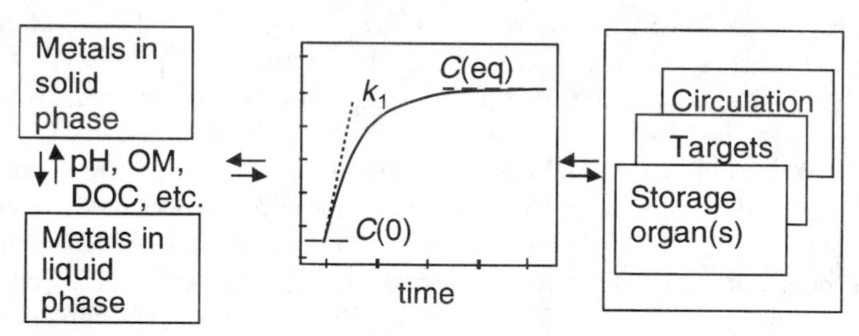

Figure 5-1 Principal processes involved in the concept of bioavailability. Environmental availability is envisaged as partitioning of heavy metals between the soil solid phase and the pore water (left). Environmental bioavailability is depicted by means of toxicokinetic uptake characteristics (uptake rate constant, k_1, and the equilibrium concentration, C(eq) [middle]). Toxicological bioavailability is shown (right) on the basis of internal recirculation and storage processes of the metals assimilated, resulting in transport to the target sites of toxic action (data from Posthuma, Notenboom et al. 1998). (DOC = dissolved organic C; OM = organic material)

Environmental availability is indicated as the physicochemical-driven distribution of metals over soil constituents. This aspect of bioavailability has already been dealt with in Chapter 2 and will not be touched upon in this chapter. Environmental bioavailability is envisaged as a dynamic process in which species- and compound-specific accumulation and excretion kinetics play central roles. Toxicological bioavailability is schematically displayed by means of redistribution and storage processes of assimilated chemicals in the body of the organism. In this chapter, the focus will be on environmental bioavailability. Toxicological bioavailability will be dealt with only briefly. Soft-bodied, soil-dwelling organisms are generally thought to be exposed mainly by the pore water, whereas additional uptake routes (such as direct uptake of the soil solid matrix or uptake via food) also may be relevant for hard-bodied, soil-dwelling organisms. Uptake routes of metals for specific taxa of soil invertebrates, therefore, are placed centrally in this chapter.

Environmental Bioavailability: Uptake Routes for Soil Invertebrates

Soil invertebrates are able to accumulate metals via a number of distinct uptake routes. Most information for arthropods concerns food uptake. The impact of soil properties on metal uptake and toxicity for Lumbricidae and Collembola has recently received increasing attention (Belfroid and van Gestel 1999). In this section, the various uptake routes will be evaluated, with emphasis on the organism- and soil-specific factors affecting metal accumulation.

Uptake via food

Microhabitat, trophic level, and food specificity are the most important species-specific characteristics leading to clear differences in availability of metals for uptake by different invertebrates exposed in the same soils, under similar conditions. According to Indeherberg et al. (1998), metal uptake via food is determined by the physicochemical properties of the gut content. Hardly any information is available on the impact of microorganisms and the physicochemical conditions of the digestive system in the gut or on metal assimilation by soil invertebrates. As stated by Hopkin (1989), it may be assumed that the processes that determine metal uptake in the gut are similar to the processes that determine sorption and desorption of metals in the soil. In addition, Hopkin stated that the rate of metal uptake via food depends on the rate of food consumption and the structure and functioning of the digestive tract. Terrestrial invertebrates show considerable discrimination in the types of food they will eat. Isopods, millipedes, and mollusks, for example, are able to detect quite subtle differences in the chemical composition of plant material (Hopkin 1989). Difficulties in determining the exact nature of the diets of terrestrial invertebrates in the field has complicated attempts to construct pathways of metal transfer between trophic levels in ecosystems. Different species of the same taxonomic group may consume completely different food. Siepel (1995) showed that the selection of food by mites is an important factor that needs to be taken into account. Metals are accumulated at high levels in fungi, especially in the cell walls. Fungi-consuming mites can be grouped in 2 classes: so-called "browsers" and "grazers." Grazers contain the enzyme chitinase in their digestive tract, enabling them to digest the cell walls of the fungi. Browsers do not contain this enzyme. This explains why lower quantities of grazers than browsers were present in a Pb-contaminated area, whereas grazing mites contained higher internal Pb levels as compared to other mite species. Characteristic differences between metal uptake by grazing and browsing mites were confirmed in a laboratory experiment carried out with the oribatide mites *Nothrus silvestris* (grazer) and *Ceratoppia bipilis* (browser). Fungi-consuming larvae do not contain chitinase either, and according to Siepel (1995), they therefore are able to limit metal uptake.

Some authors have suggested that terrestrial invertebrates may be able to select food containing optimum amounts of essential metals. This would require the presence of feedback mechanisms controlling feeding behavior. Although such a concept is difficult to accept because there is no physiological or biochemical evidence for the presence of such feedback mechanisms controlling metal regulation in any organism, Dallinger (1977) and Dallinger and Wieser (1977) showed that the isopod *Porcellio scaber* was able to discriminate between 3 batches of leaves, which differed only in their Cu concentrations. As stated by Hopkin (1989), extreme care should be taken in interpreting the results of such experiments. It is possible, for instance, that the isopods were not discriminating on the basis of Cu levels in the food but that the organisms responded to different levels of the anion, which they could probably taste. In addition, application of metals as salts to leaf material also may change the texture and taste of the food by killing microorganisms. Because invertebrates may simply reject food that tastes "unpleasant," discrimination against food contaminated heavily with metals is easier to explain. On the other hand, some invertebrates in metal-polluted sites may not be exposed to higher concentrations of metals because their diet is not contaminated. Thus, spiders may survive in heavily contaminated areas, if they feed on flying insects that have migrated in from an adjacent uncontaminated area.

A clearly different way of metal uptake is observed for snails. Snails appear to have a high efficiency of metal assimilation from food. In addition, they are able to accumulate metals via the foot mucus (Hopkin 1989).

The rate of food consumption is dictated by the concentration of the most limiting essential nutrient in the diet. As a consequence, many other substances may be ingested in amounts far in excess of requirements, necessitating regulation of their uptake, transport, and excretion. Differences in feeding rates and assimilation efficiency will have a profound effect on assimilation of metals by terrestrial invertebrates. When food of "high quality" is available, often a strategy of feeding as fast as possible is adopted because relatively little energy is required to solubilize the initial fractions of essential nutrients released from the food. When food of "poor quality" predominates, feeding rates drop because less energy is required to retain food in the gut and allow digestive enzymes to work on it longer, than to search for food in which the nutrients are more available (Hubbell et al. 1965). In addition, it is well established that rates of assimilation and toxicity of metals in animals are strongly related to the ligands to which the elements are bound in the food. The differences relate to the ease with which metals are converted to an "available" form in the lumen of the gut. The ratios of different metals in the food are also important. Metals may compete for the same routes of uptake (or share the same uptake sites in the cell membranes), leading to antagonistic effects, or the presence of one metal may stimulate the uptake or effects of other metals (synergistic metals).

In summary, the information provided above clearly shows that food uptake of heavy metals by terrestrial invertebrates is affected mainly by various biological species-specific factors. Quantification of the contribution of metal assimilation via food, however, has not been carried out yet.

Comparison of uptake via soil and uptake via food

Effect concentrations (ECs) may be expressed in both external and internal concentrations. Crommentuijn et al. (1994, 1997) studied Cd uptake in the springtail *Folsomia candida*. The toxicity of Cd to *F. candida* in artificial soil was compared to toxic levels obtained by direct exposure via the food. They found that much higher Cd ($CdCl_2$) levels in the food were needed to obtain similar effects as in the case of exposure solely via the artificial soil. However, differences in external ECs became significantly lower following normalization of the external Cd concentrations upon the organic matter content of the substrate (10% for soil, 100% for food).

Crommentuijn et al. (1997) observed that internal Cd concentrations equilibrate after about 20 days of exposure in artificial soil. The $EC50_{growth}$ of *F. candida*, when expressed on the basis of external Cd concentrations, was found to be equal to about 800 mg Cd/kg dry soil. The internal $EC50_{growth}$ was estimated to be 200 to 250 mg Cd/kg dry weight (dw). For Cd exposure via food, Gambrell (1993) found that internal Cd concentrations equilibrate after 15 to 22 days of exposure. Again, much higher Cd concentrations in food were necessary to obtain the same effects ($EC50_{growth}$ 3184 mg Cd/kg food). This external effect level was shown to correspond to a level of 219 mg Cd/kg dw, which is similar to the value reported by Crommentuijn et al. These results clearly show that exposure via soil of *F. candida* is far more effective than exposure via food, whereas the resulting internal concentrations are independent of the uptake route.

Similar conclusions may be drawn from the results reported by Smit (1997). Smit determined Zn toxicity for the springtail *F. candida* in a field soil amended with $ZnCl_2$ and allowed to age in either the laboratory or the field. In addition, Zn toxicity via food was determined. Effect concentrations based on either total metal content of the substrate or on the water-extractable Zn content of the soil differed by several orders of magnitude. However, they became similar when toxicity was expressed on the basis of the organic matter content of the substrate (food or soil). Also, internal effect levels upon exposure via food (on average 387 mg Zn/kg dw) were similar to the internal effect levels of 118 to 452 mg Zn/kg dw following direct exposure via the soil. On the other hand, Smit and van Gestel (1996) failed to explain differences in Zn toxicity in *F. candida* between soils on the basis of internal Zn concentrations.

Zonneveld (1997) and Quigney (1998) carried out exposure experiments with the isopod *P. scaber*. In these experiments, the animals were exposed to Cd or Zn via food, soil, or a combination of both substrates. Food and soils were spiked with metal chloride salts of the metals mentioned. Exposure of the isopods to similar metal concentrations in food and soil led to higher internal metal levels in the case of soil exposure, which was similar to the findings of Crommentuijn et al. (1994) depicted previously. For Zn, the lethal concentration to 50% of organisms (LC50) was found to be 1421 mg/kg dry soil. In contrast, only 32% mortality was observed following exposure to food containing 5365 mg Zn/kg dry food. No dose-related effects of Cd were observed in these studies. Metal uptake via soil and food was found to be additive; higher amounts of metal were accumulated following simultaneous exposure via both soil and food, as compared to exposure to either soil or food separately. When exposed to Cd via the soil, the isopods are capable of Cd storage in the remainder of the body, after which the Cd is transported to the hepatopancreas. On the other hand, Cd accumulated from the food is stored directly in the hepatopancreas. For both exposure routes, a shift in time of the Cd burden towards the hepatopancreas takes place (Zonneveld 1997).

Descamps et al. (1996) exposed the millipede *Lithobius forficulus* to Pb. Internal Pb levels were shown to increase quickly following exposure via food, after which they leveled off. Similar findings were observed for Cd. This typical behavior may be explained by the development of an active excretion system upon chronic Pb or Cd food exposure. Internal Pb concentrations in animals exposed to contaminated soil increased quickly to reach a plateau value of 50 mg Pb/kg dw within 1 week of exposure. On the other hand, animals exposed to 219 mg Pb/kg food reached a maximum internal Pb concentration of 136 mg Pb/kg dw, after which this internal concentration decreased to a steady-state concentration of about 28 mg Pb/kg dw. These results suggest that millipedes are capable of accumulating Pb via the integument; food is not the sole route of uptake for these hard-bodied organisms, and porewater-mediated uptake may be an important exposure route.

In summary, from the information provided above, it may be deduced that many soil-dwelling organisms have a number of uptake routes for heavy metals. Often, exposure linked to the soil solid phase contributes most to the internal metal levels observed, as compared to uptake via food. It should be realized, however, that in addition to species-specific uptake routes, various storage mechanisms within specific taxa may reduce the truly bioavailable metal fraction present within the organism (i.e., the toxicological bioavailable fraction). In addition, as shown, for instance, by Siepel (1995), Chen and Mayer (1998), and Selck et al. (1998), factors such as the selection of food, the speciation of the metals in the food, the feeding behavior of the animals, the strategies for avoiding uptake of contaminated food, and the impact of enzymes and amino acids on metal speciation in the gut are determining factors in actual metal uptake by invertebrates. At present, insufficient information is available to make a clear, quantitative distinction between the various modes of uptake, let alone to quantify toxicological bioavailability.

Differences in Accumulation Levels among Taxonomic Groups

Heikens et al. (2001) compared differences in bioaccumulation of Cd, Cu, Pb, and Zn in 9 taxonomic groups exposed in soils containing different metal concentrations. With the exception of snails, the taxonomic groups selected represent the most important terrestrial invertebrate species. A metal-dependent increase in body concentrations with increasing external metal concentrations was observed for most taxonomic groups. The overall pattern of internal metal concentrations was Pb > Cd > Cu, Zn. This might reflect active regulation of internal body levels, among other things. Accumulation levels among taxonomic groups differ significantly by factors between 2 and 12. Overall, metal concentrations were high in the Isopoda and low in Coleoptera. Lumbricidae (a group of organisms commonly used in risk assessment) had average internal levels of Cu, Zn, and Pb. As reported by Hopkin (1989), Cd levels in Lumbricidae were relatively high. Heikens et al. (2001) concluded that it is necessary to include a detailed feeding behavior scheme in the risk assessment of heavy metals for invertebrate-eating animals. It should be noted that in this study, the impact of differences in soil composition was not explicitly taken into account. A proper comparison of differences in accumulation would require either the inclusion of the impact of soil composition and exposure routes or the derivation of data from the same sites; the latter is illustrated in Table 5-1 and Figure 5-2. Heikens et al. (2001) derived regression equations describing the relationship between internal metal concentrations in Lumbricidae and total metal concentrations in the bulk soil. The results obtained for Pb are depicted in Table 5-1 and Figure 5-2. Data of Ma et al. (1983), Corp and Morgan (1991), and Neuhauser et al. (1995), and data reported in the study of Heikens et al. (2001), provided the basis for the regression equations shown.

Table 5-1 Coefficients for monovariate regression formula describing the quantitative relationship between log-transformed body concentrations of Pb in Lumbricidae and total Pb concentrations in the bulk soil: $\log[Pb]_{Lumbricidae} = \log a + b \cdot \log[Pb]_{soil}$. (Units of $[Pb]_{Lumbricidae}$ and $[Pb]_{soil}$ are μg/g dw.)

Equation	Log *a*	Log *b*	Reference
1	0.30	1.00	Ma et al. 1983
2	0.10	0.69	Corp and Morgan 1991
3	0.10	0.74	Neuhauser et al. 1995
4	0.80	0.64	Corp and Morgan 1991
5	0.00	0.61	Neuhauser et al. 1995
6	–0.40	0.62	Heikens et al. 2000

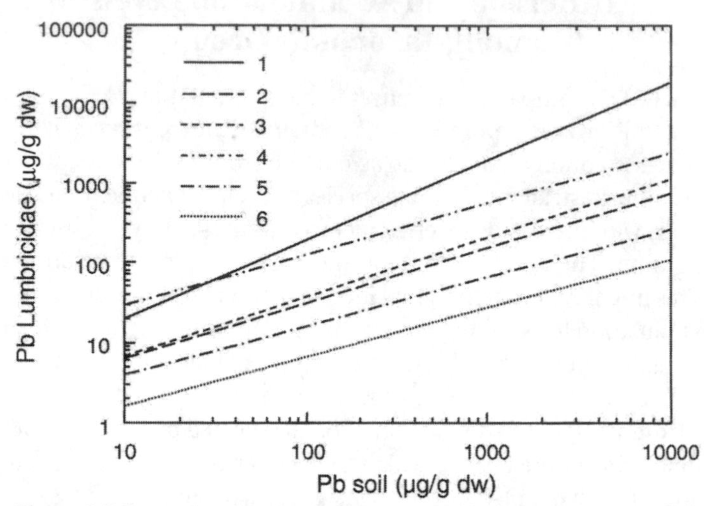

Figure 5-2 Regression equations relating internal body concentration of Pb in Lumbricidae to total Pb concentrations in the solid soil matrix. The lines are arranged according to the numbering used in Table 5-1.

From Figure 5-2, it may be deduced that internal Pb levels are related to the total Pb concentrations in the bulk soil (for soils with limited variance in soil properties, data are not shown). However, when uptakes are compared in structurally different soils, significant differences in internal Pb concentrations become apparent. In view of the likelihood of total Pb concentrations in soils with similar physicochemical properties being confounded to, for instance, Pb concentrations in the soil solution, these findings do not provide conclusive evidence regarding the dominant metal uptake routes for Lumbricidae.

A comparison of Cd and Pb levels in a number of soil invertebrates showed that there are large differences in internal metal among invertebrate species. The differences are related to the feeding behavior of the organisms. Saprofagous organisms (such as earthworms and isopods) contain the highest concentrations (Weigmann 1989). According to Weigmann, direct contact with soil and pore water might explain this observation for earthworms, while additional metal uptake via the food might be responsible for the isopods observations. The other organisms investigated all have a predatory feeding habit and consequently accumulate less metal. These findings complement the results of Laskowski (1991) and Janssen et al. (1993), who found that metals do not accumulate strongly within the terrestrial food chain. Cd and Pb are not transferred from the quarry to the predator, and direct uptake from either the soil or the litter is less important for these predators.

Roth (1992) showed that spiders, oribatide mites, centipedes, fly larvae, and beetles can accumulate Cu at much higher levels than do Lumbricidae and enchytraeid worms. This shows that "hard-bodied" organisms, too, are capable of accumulating large quantities of metal. Because this observation is also valid for predators, these findings show that, apart from food uptake, additional uptake routes are of importance. From results reported by Carter (1983) and Scharenberg and Ebeling (1996), it may be deduced that organisms living in direct contact with the soil matrix and/or feeding with depository material (earthworms, enchytraeid worms, centipedes, snails, isopods) contain the highest metal contents. Predators such as spiders and beetles usually contain the lowest levels, whereas oribatide mites and springtails have metal levels in between these extremes. Ground beetles are an exception to this generalization because the levels of Cu and Zn in these organisms are highest, while their Cd levels are relatively high. Zn levels do not differ much among taxonomic groups, which might be a consequence of active regulation of this essential element. It is striking that a difference in Zn and Cd levels between the adult and larval stadia of the ground beetle *Pterostichus melenarius* was observed by Carter (1983). This suggests that beetles in their adult stadia are less exposed than beetles in their larval stadium, which probably is due to a difference in their mode of life and their architecture; organisms in a larval stadium have a softer skin and reside mainly in the soil, whereas adult stadia have a harder skin and usually reside on the soil surface. In addition, it is reported by Janssen et al. (1991) that some adult beetles are able to excrete metals, which will contribute to lower levels in the adult stadium.

In summary, it is clear that internal metal concentrations may vary considerably among terrestrial invertebrate species. Most data are available for Cd, Cu, Pb, and Zn. In general, organisms living in direct contact with the soil matrix and/or feeding with depository material contain the highest metal contents. Predators usually contain the lowest levels, whereas oribatide mites and springtails have metal levels between these extremes. Ground beetles are an exception because the Cu and Zn levels in these organisms are highest, with high Cd levels as well. Zn levels do not differ much among taxonomic groups, which might be a consequence of active regulation of this essential element. The general picture is clouded by additional factors that affect metal uptake, such as mode of life, architecture, life stage, and the possibility of metal excretion.

Impact of Soil Properties on Metal Uptake Versus the Importance of Porewater Uptake

In general, Zn toxicity to *F. candida* decreases in soils that either have been aged under natural conditions or have been flushed with water after spiking with Zn^{2+} salts to remove excess counterion. Flushing, however, also may cause other alterations, such as removal of components of organic matter. In these soils, the internal ECs for reproduction are about 2.0 to 2.5× higher than in fresh (untreated) soils. Smit (1997) interpreted these results as showing the impact of the chloride anion on internal effect levels. Effect concentrations in a range of soils were found to correlate best with water-soluble metal concentrations. Similar results were found by

Posthuma, van Getsel et al. (1998) for Zn toxicity for earthworms (*Eisenia andrei*), enchytraeid worms (*Enchytraeus crypticus*), and plants (*Trifolium pratense*). For all species mentioned, differences in toxicity between freshly treated soils and soils aged under natural conditions are reduced when effects are related to either water-soluble or 0.01M $CaCl_2$-extractable concentrations. However, this conflicts with the data given by Smit and van Gestel (1996), as referenced above. It must be stressed again that internal ECs for these organisms are independent of the soil type and the history of metal pollution.

On the other hand, it is reported by Crommentuijn et al. (1997) that external ECs of Cd (freshly spiked as $CdCl_2$ to a number of artificial soil media containing varying amounts of organic matter and a range of pH values) for *F. candida* diverge when expressed on the basis of external water-soluble concentrations. Internal Cd levels in the springtails, however, were shown to correlate strongly to water-soluble Cd concentrations. According to Crommentuijn et al. (1997), these findings show that although uptake may be explained on the basis of water-soluble Cd concentrations, the sensitivity of springtails to Cd also is determined (in part) by soil properties such as pH and organic matter content. Similar results have been reported by Vonk et al. (1996) on Cd toxicity for microorganisms, springtails, and earthworms in 3 soils at 3 pH values.

Janssen, Posthuma et al. (1997) studied the impact of soil properties on metal uptake by earthworms (*E. andrei*) following exposure during a period of 3 weeks in 20 natural field soils. These authors found that the same soil properties affecting the metal partitioning between the soil solid phase and the pore water were also the dominant soil properties affecting metal accumulation in the worms. Soil pH was the most dominant soil property in this respect. Although direct uptake via the gut wall cannot be completely ruled out, these findings suggest that metal uptake by earthworms takes place predominantly via the pore water or via an uptake route that is related to porewater uptake. It cannot be ruled out that confounding factors such as increased Al toxicity at low pH and direct effects of pH on metal binding to the cell walls may have profoundly impacted the findings reported.

Janssen et al. (1996) related uptake of Cs by the Lumbricidae *E. andrei* and *Lumbricus rubellus* from a freshly treated sandy soil to the total Cs content of the soil and the porewater concentration. In contrast to the findings mentioned in the preceding paragraph, these authors conclude that for both worm species investigated, apart from possible uptake via the pore water, additional uptake via the food may be of importance. But those findings might be related to the particular physicochemical properties of Cs.

With regard to porewater-related uptake, it should be noted that available information shows that organisms are not exposed directly to the bulk soil solids. Instead, metal uptake tends to proceed via the soil solution, which in turn may be actively modified by the organism. For example, terrestrial invertebrates possess a wide range of digestive enzymes. These enzymes are most active under specific chemical conditions, the most important of which is pH. Digestive fluids are therefore often buffered to maintain the pH within a certain range. As reported by Hartenstein (1964), Humbert (1974), and Wallwork (1983), the pH of the gut of Isopods, Collembola, earthworms, and most other detrivores is usually within 1 unit of neutrality. The common observation of increased soil pH during toxicity testing of these species can be explained by active buffering of the pH of ingested soil material. In some species, the pH of particular regions of the gut may be strongly acidic or alkaline to promote degradation of particular fractions of the food. Especially at low pH, this will strongly increase metal solubility.

Weeks and Rainbow (1993) determined uptake of Cu and Zn by 2 talitride amphipods from food and compared this mode of uptake with results obtained in earlier studies in which metal uptake in water was measured. *Orchestia gammarellus* is adapted to living on the land, while *Orchestia mediterranea* lives in the aqueous environment. Cu uptake from the food was the dominant uptake route for *O. gammarellus*, whereas for *O. mediterranea*, additional uptake via the gills was needed to fulfill the need for the essential element Cu. Zinc was taken up by both species predominantly via the food. The results of this study show that the routes of uptake may vary for some species and may depend, for instance, on the necessity of controlling the internal levels of essential elements. In addition, studies of Selck et al. (1998) have shown that Cd uptake by the polychaete *Capitella* sp. I is dependent on the type of organic matter present in the solid phase. Cd uptake increases with increasing amounts of well digestible exopolymers. In part, these findings may be explained by the observation that exopolymers may strongly increase the availability of metals. The finding that Cd uptake is increased upon increased residence time of the sediment in the gut, however, shows that digestion of the exopolymers is needed to release the Cd present. Hence, metal uptake is dependent on the binding capacity of the solid phase and both the amount and the physicochemical properties of organic material.

With regard to free metal uptake via the pore water, it was recently stressed by Plette et al. (1999) that, in practice, it can be very difficult or even impossible to measure metal ion sorption to biota in a soil system. In terrestrial oxic systems, the pH and the Ca concentration (as a competitor for the same sorption sites as the metal ion) are considered to be two of the most important factors that affect the distribution of the metal over the components (including biota) present in the soil system. In principle, a model for metal ion binding should be able to predict biosorption for a wide range of conditions with respect to pH, solution composition, and ionic strength. According to Plette et al. (1999), a pH-dependent Freundlich model is the simplest model that can describe the metal ion binding to biota, the dissolved organic matter (DOM), and the soil solid phase. Model calculations show that pH and composition of the pore water can have a large impact on metal uptake. It is shown that the net effect of pH on biosorption is dependent on the characteristics of both the soil–porewater system and the organism and can be completely reversed from one system to another. As an example, in a sandy soil, Cu availability to maize roots was strongly reduced upon increasing pH, whereas for yeast, an increase of pH in solution led to enhanced uptake and enhanced effects.

In summary, metal uptake takes place via a complex interplay of organism-specific, soil-specific, and porewater-specific factors, which may be affected additionally by complex factors such as the necessity of retaining internal concentrations of essential elements at a specific level. Actually, routes of uptake for a given species may even vary for different metals. It may be concluded that assessing the dominant route of uptake for a specific organism often is difficult. Nevertheless, the available information shows that, in general, organisms are not exposed directly to the bulk soil. Instead, metal uptake tends to proceed via the (modified) soil solution. Taking the modifying factors mentioned in the preceding paragraph into account, metal uptake can be best described on the basis of the composition of the soil solution, which may be actively modified by the organism.

Impact of Soil Type on the Dynamic Aspects of Bioavailability: Accumulation and Excretion Kinetics

In the previous sections, the focus was on uptake routes and parameters affecting uptake, including the impact of biotic and abiotic factors. In this section, attention will be paid to the dynamic aspects of bioavailability, that is, the accumulation and assimilation kinetics of individual metals. Body residues often are better estimates of the amount of a chemical at the sites of toxic action in an organism than are ambient soil concentrations because bioavailability differences among soils are explicitly taken into account when body residues are considered. Often, however, insufficient attention is paid to the rate and extent at which tissue concentrations respond to soil concentrations and soil characteristics. The body residue concept is based on the assumption that the toxicant level in the organism must reach some threshold value at the target site (e.g., cell membranes for narcotic agents) before effects start to occur. Once taken up, metals may be partitioned into either biologically available, biologically unavailable, or storage fractions (Figure 5-1). Biologically available metals, in their turn, can participate in essential metabolic functions or, in the case of nonessential elements or excess essential metals, can contribute to toxicity. It should be noted that, even when there is uptake, there is not necessarily toxicity because the organisms may sequester metals and thus avoid physiological impact. In addition, organisms may change their environment, thereby altering the metal distribution in soil and inside the organism (e.g., metallothionein). Therefore in some species, external (available) metal concentrations, rather than internal (noncomplexed) concentrations, may directly determine toxic effects.

For soft-bodied organisms (such as protozoa, nematodes, Lumbricidae, enchytraeid worms, some insect larvae), it usually is assumed that uptake is governed — at least for a major part — by transport via (pore) water, and it is the free metal ion in water that often is considered to be the toxic species that can actually be taken up. For hard-bodied organisms (arthropods such as spiders, harvestmen, mites, insects, centipedes, millipedes, Isopods, and some crustaceans living on the soil), additional uptake routes such as ingestion of soil and food intake may contribute significantly to the metal burden in the organism and hence to the toxic effects observed.

Kinetic studies can be used to predict the physiological fates of pollutants in exposed organisms, and a number of studies have used kinetic models to describe the accumulation of nonessential metals (Janssen et al. 1991) and hydrophobic organic chemicals (Belfroid et al. 1995) by soil invertebrates. Similar approaches were used by Walker et al. (1996), Peijnenburg, Baerselman et al. (1999), Peijnenburg, Posthuma et al. (1999), and Spurgeon and Hopkin (1999). Although most biological testing of soils is typically carried out with earthworms, and despite the plethora of data detailing metal concentrations in earthworms collected from contaminated soils, relatively little is known about the accumulation and excretion kinetics of individual metals. Almost no kinetic information for other invertebrates has been reported.

Spurgeon and Hopkin (1999) assessed the accumulation and excretion kinetics of Cd, Cu, Pb, and Zn in time-series studies with the earthworm *Eisenia fetida* in 2 contaminated field soils, collected from sites around smelting works, and an uncontaminated soil. For worms exposed to the clean soils, no consistent trends of uptake were found for any of the metals examined. The 4 metals were generally accumulated in the 2 polluted soils. Different time-dependent patterns of uptake were found for essential and nonessential elements. Equilibrium was reached within 7 days of exposure for the essential elements Cu and Zn. Body burdens failed to reach equilibrium within 40 days of exposure for the nonessential metals Cd and Pb. A one-compartment model was fitted to the data to model metal accumulation and allow accumulation and excretion rates to be calculated. Accumulation rates and equilibrium concentrations were highest in the most polluted soil, indicating that uptake was related to the total soil metal concentration. However, because these authors experimentally determined only a limited number of soil properties, it cannot be excluded that this observation covaries with other modes of uptake. Also, differences were found in the accumulation rates of the selected metals. Generally, values were low for Cd, intermediate for Cu and Pb, and high for Zn. These findings are similar to those of Neuhauser et al. (1995).

Equilibrium concentrations could not be calculated for Cd because of the fact that the calculated elimination rate was effectively 0 in the field soils, indicating that Cd body burdens at a given soil concentration will be dependent on the duration of exposure. In contrast, rapid rates of excretion were found in all cases for the essential metals Cu and Zn. Rapid rates of loss of Cu and Zn have also been found in other studies with invertebrate species (Hopkin 1989; Dallinger 1993), although Cu excretion rates may be lower for hemocyanin-dependent groups such as molluscs and isopods (Hopkin 1989). The levels of Cu and Zn stabilized at concentrations comparable to worms in the control soil and showed no further fluctuation over the excretion period.

The differences found in the excretion kinetics of essential and xenobiotic metals in *E. fetida* suggest contrasting physiological fates for these metals. On the basis of additional excretion studies (showing either slow excretion or absence of excretion), it was concluded that it is probable that the main detoxification pathways of Cd and Pb are sequestration within inorganic matrices or binding to organic ligands. The presence of such storage detoxification mechanisms for Cd and Pb in a number of earthworm species has been demonstrated before (Cancio et al. 1995; Morgan et al. 1995). By comparing rates of accumulation and excretion, Spurgeon and Hopkin (1999) were able to demonstrate that it is the difference in excretion, instead of the difference in accumulation, which underlies the higher equilibrium concentrations.

The uptake kinetics of As, Cd, Cr, Cu, Ni, Pb, and Zn in the oligochaete worms *Enc. crypticus* and *E. andrei* were studied in 20 moderately contaminated field soils by Peijnenburg, Baerselman et al. (1999) and Peijnenburg, Posthuma et al. (1999). Total metal concentrations typically varied by more than 2 orders of magnitude. The objective of the research was to quantify the modulation of bioavailability of the metals as a function of soil characteristics. The working hypothesis was that interactions in soil modulate the distribution of metal species over the soil phases, which, in combination with organism-specific factors, modulate metal uptake. Body residues were determined at preset time intervals during a maximum period of 35 (*Enc. crypticus*) or 63 (*E. andrei*) days of exposure. Apart from possible toxic effects related to changes of body weight, toxicological bioavailability was not addressed. Compartment modeling was used to yield data on the uptake dynamics of the process and the equilibrium status. At equilibrium, bioaccumulation factors (BAFs) were calculated relative to soil subcompartments. A BAF represents the ratio of the internal concentration of a chemical in an exposed biological system to the concentration in the exposure medium. In aquatic accumulation studies, it is customary to use the total concentration in the water phase to calculate BAFs. In aquatic studies, it is implicitly assumed that the uptake rate constant is independent of the bioavailable metal concentration in the aqueous phase and is modified solely by system-related factors such as variation in pH, mineral composition, and suspended solids. Fortunately, because total metal concentrations and metal concentrations that are available for uptake by aquatic organisms usually are highly correlated, the implicit assumption usually is met in aquatic studies. Similarly, in terrestrial accumulation studies, it is common to use the total soil concentration for the purpose of calculating BAFs, usually yielding biota-to-soil accumulation factors (B_SAFs) (Dowdy and McKone 1997). However, as indicated in "Impact of Soil Properties on Metal Uptake Versus the Importance of Porewater Uptake" (p 99), this procedure fully ignores the facts that soil organisms do not directly respond to total concentrations and that soil organisms potentially have different routes of uptake. Therefore, not only total metal concentrations in the soil compartment but also total metal concentrations in the pore water, calculated free metal ion activity in the pore water, and extractable metal concentrations were used by Peijnenburg, Baerselman et al. (1999) and Peijnenburg, Posthuma et al. (1999) to calculate BAFs and uptake rate constants. Extraction with 0.01 M $CaCl_2$ was considered as an operationally defined method for quantifying bioavailability. The following generalized formula was used to analyze the dynamic change of body concentration in the worms over time:

$$\frac{dC_w}{dt} = k_1(x)C_x - k_2C_w \qquad \text{(Equation 5-1)},$$

where t is time (days) and C_w is metal concentration in the worms (mmol/kg dw_w). The subscript x refers to the hypothesized bioavailable phase in the exposure medium (soil [s] or pore water [pw]) so that C_x is either

- the total metal concentration in the soil compartment (mmol/kg_s), C_s;
- the total metal concentration in the pore water (mmol/L_{pw}), C_{pw};
- the 0.01 M $CaCl_2$-extractable metal concentration (mmol/kg_s), C_{CaCl_2}; or
- the calculated metal ion activity in the pore water (mmol/L), (Me^{2+}).

The general solution of Equation 5-1 is

$$C_w(t) = C_w(0)e^{-k_2t} + \frac{k_1(x)C_x}{k_2}(1 - e^{-k_2t}) \qquad \text{(Equation 5-2)},$$

where k_1(x) is the uptake rate constant (either kg_x/kg $dw_w \times$ day, or L_x/kg $dw_w \times$ day), and k_2 is the elimination rate constant (1/day). Equation 5-2 is derived assuming that (C_x) is not changing during the uptake period. Multivariate expressions were derived that relate steady-state internal body concentrations, BAFs, and uptake and elimination rate constants to easily determinable soil characteristics.

It was found that Cu concentrations in *Enc. crypticus* did not differ significantly from those in the culture, which apparently points to regulation of the internal body concentrations. Accumulation of Cd and Zn was observed in all soils tested. However, for most soils, no equilibration of the internal Cd concentrations was achieved. Thus the data suggest that, for most soils, uptake of Cd is still in its linear phase. Accumulation of Pb occurred in 9 of the 20 soils. No detectable Pb levels could be found in the organisms that were exposed in the remaining soils, even after 35 days of exposure. Extraction of the soils with a 0.01 M $CaCl_2$ solution (Peijnenburg, Baerselman et al. 1999) discriminated between soils in which Pb accumulation by *Enc. crypticus* does take place and soils for which this is not the case. For those soils in which the 0.01 M $CaCl_2$ extraction yielded amounts of Pb above the detection limit for the extracts of 0.1 μmol/kg, there is also accumulation of Pb in the organisms, as opposed to the soils where Pb levels in the extracts were not detectable. Figure 5-3 shows some characteristic accumulation curves for each of the metals studied.

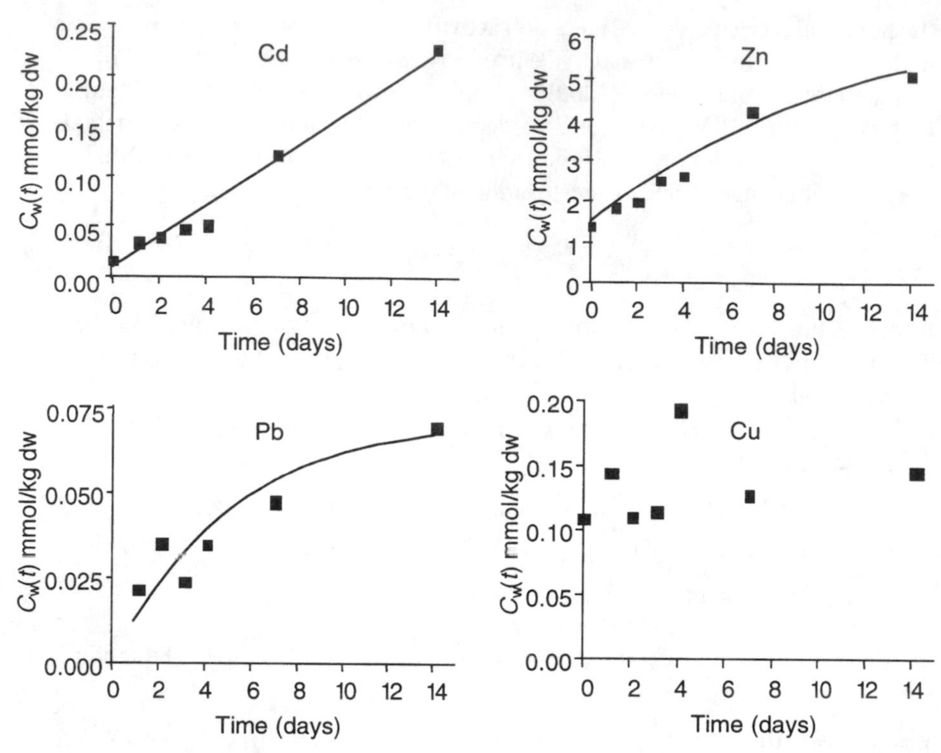

Figure 5-3 Characteristic observations and accumulation curves of Cd, Zn, Pb, and Cu in *Enc. crypticus* following exposure in Dutch field soils. Observations are averages of 4 replicates (data from Peijnenburg, Posthuma et al. 1999).

For *E. andrei*, 2 characteristic types of uptake could be distinguished:

1) Uptake following one-compartment behavior. In a number of cases, the rate of elimination was found to be close to 0, which resulted in apparent linear uptake of the metal over the whole exposure duration.
2) Fast accumulation, leading to internal concentrations that were usually at steady state within 1 to 3 days of exposure.

In all soils studied, fast accumulation of type 2 was the sole uptake pattern observed for Cu and Zn, 2 essential metals whose body concentrations are known to be regulated. Thus, whereas total external Cu and Zn concentrations varied by as much as a factor of 100 and 590 respectively, steady-state body concentrations of Cu and Zn were found to vary by no more than a factor of 5, with an almost immediate response of the body concentrations. Similar to Cu and Zn, fast accumulation of Cr and Ni at levels exceeding the body concentrations in the culture was typically observed. Log-transformed internal steady-state concentrations are plotted against log-transformed total metal concentrations in Figure 5-4.

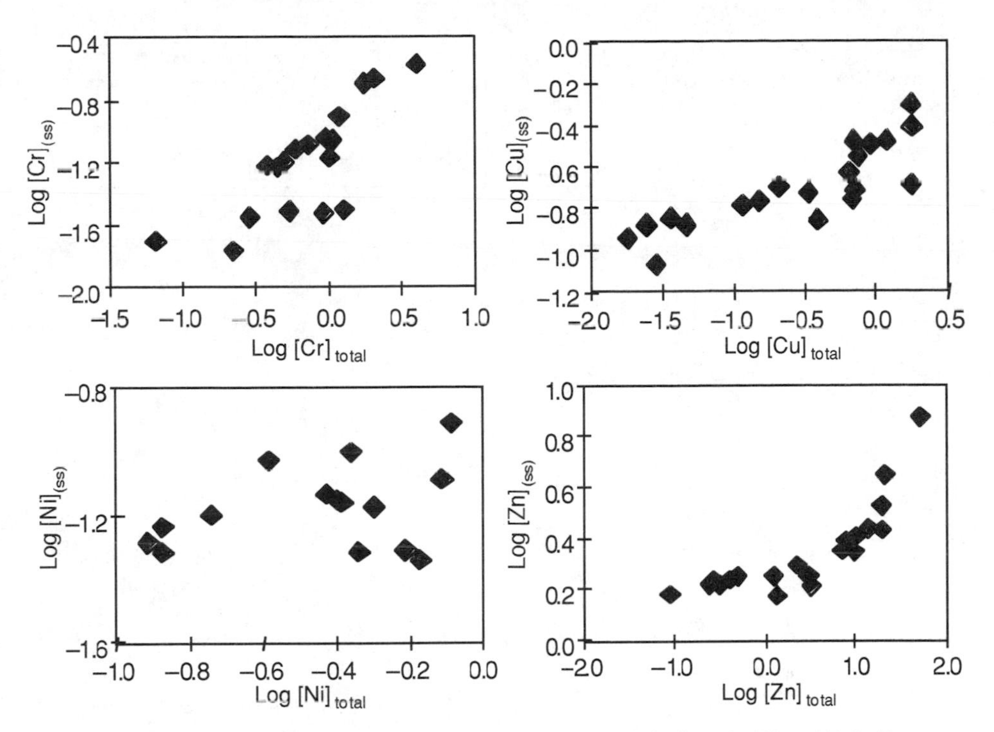

Figure 5-4 Log-transformed internal steady-state concentrations (ss) of Cr, Cu, Ni, and Zn in *E. andrei* (mmol/kg dw_w) after exposure in 20 Dutch fields versus log-transformed total metal concentrations (mmol/kgs) (data from Peijnenburg, Baerselman et al. 1999)

The graphs presented in Figure 5-4 suggest that steady-state concentrations of Cr and Cu are related to total metal concentrations, whereas for Ni the data appear to be scattered. It should be noted that, in the case of Ni, the variance in steady-state body concentrations is limited. It also may be deduced from Figure 5-4 that steady-state Zn levels are independent of external Zn concentrations up to 3 $mmol/kg_S$, irrespective of the soil characteristics. Above this level, log-transformed internal concentrations tend to increase linearly with log-transformed total Zn levels. It should be noted that the steady-state Zn concentration (on average: 1.7 mmol/kg dw) that is found in the soils containing Zn in concentrations below 3 mmol/kg corresponds well to the constant level reported by van Gestel et al. (1993). Monovariate regression formulas describing the quantitative relationship between log-transformed steady-state body concentrations of Cr, Cu, and Zn for *E. andrei* after exposure in Dutch field soils and the log-transformed total metal content are given in Table 5-2.

Arsenic accumulation in general adhered to the toxicokinetic model. No accumulation of As was found in field soils having both a pH ($CaCl_2$) value of less than 6 and total As levels below 0.4 mmol/kg. However, at pH values exceeding 6.75, elimina-

Table 5-2 Monovariate regression formulas describing the quantitative relationship between log-transformed steady-state (ss) body concentrations (C_w(ss)) of Cr, Cu, Ni, and Zn for *E. andrei* after exposure in Dutch field soils and the log-transformed total metal content of soil[a]

Metal	Equation[b]	Statistics	Requirements
Cr	$\log C_w(ss) = -1.05 + 0.69 \cdot \log[Cr]_s$	$R^2_{adj} = 0.61$ $n = 18$ s.e. = 0.22 $F = 28.2$ $P < 0.001$	Uptake only if $[Cr]_s > 0.2$ mmol/kg
Cu	$\log C_w(ss) = -0.54 + 0.25 \cdot \log[Cu]_s$	$R^2_{adj} = 0.69$ $n = 20$ s.e. = 0.11 $F = 43.3$ $P < 0.001$	None
Ni	No significant correlation	—	Uptake only if $[Ni]_s > 0.069$ mmol/kg
Zn	$\log C_w(ss) = 0.42 + 1.45 \cdot \log[Zn]_s$	$R^2_{adj} = 0.83$ $n = 9$ s.e. = 0.11 $F = 39.0$ $P < 0.001$	$[Zn]_s > 3$ mmol/kg

[a] Data from Peijnenberg, Posthuma et al. 1999.

[b] Total concentration of metal in soil; ss indicates steady state, and C_w(ss) indicates body concentrations.

tion is virtually equal to 0, and linearly increasing internal concentrations were observed.

No significant Cd uptake by *E. andrei* was observed for soils in which the total Cd content was less than 0.01 mmol/kg. Uptake according to the one-compartment model was the general observation for the remaining soils, although in some soils, linearly increasing Cd levels (excretion rate, $k_2 = 0$) were detected over the 63-day exposure period.

Although in some cases there appeared to be a tendency toward slowly decreasing Pb levels, the common uptake pattern in 8 of the 20 soils was fast accumulation of Pb to reach internal steady-state levels that varied between 0.008 and 0.49 mmol/kg dw. Linearly increasing Pb levels (no elimination) were observed in 6 field soils, whereas the internal Pb concentrations in *E. andrei* fully adhered to the one-compartment model in the remaining 7 field soils. There were no soil-related parameters that could be used to discriminate soils in which fast accumulation of Pb takes place from soils; this result fit the slower one-compartment model. The only relevant observation in this respect was that, for all soils with pH ($CaCl_2$) below 3.9, linearly increasing internal Pb concentrations over the whole 63-day exposure period were observed.

With regard to elimination of As, Cd, and Pb, it was observed that elimination can take place only in soils for which pH ($CaCl_2$) exceeds a value of 5.

Multivariate expressions were derived to describe uptake rate constants, steady-state concentrations, and BAFs as a function of soil characteristics. As an illustrative example, the regression formulas obtained for accumulation of Cd in both earthworm species are given in Tables 5-3 and 5-4. For the enchytraeid *Enc. crypticus*, in general the most significant correlations between uptake parameters on the one hand and soil and porewater characteristics on the other hand were obtained when using the total soil concentration. The soil parameter that generally contributed most to explaining the variance among uptake rate constants and BAFs was pH (Cd, Zn), but CEC (Pb) and clay (Cd) content were also factors. Addition of Al_{ox} (aluminum [oxy] hydroxide) or organic matter slightly, but nevertheless significantly, improved the correlations obtained (for Zn and Pb). In part, these findings may be explained by covariance of total metal concentrations with these soil characteristics. For *E. andrei*, soil pH was again the dominant soil property that contributed most to explaining the variance in uptake rate constants and BAFs. For As and Pb, the formula for predicting uptake rate constants was significantly enhanced by addition of either the clay content of the soils or the dissolved organic C (DOC) content of the pore water as a descriptor. Steady-state concentrations of As in *E. andrei* can be well predicted on the basis of the clay content of the soils and the DOC content of the pore water, although the number of data is limited. For Cd and Pb, it was found that the steady-state concentrations were best described by the amount of $CaCl_2$-extractable Cd in the soil and the calculated Pb activity in the pore water. Addition of $[Ca^{2+}]_{pw}$ as a descriptor significantly improved the correlations found for Cd and Pb.

The impact of soil characteristics is grossly similar for both organisms, which suggests similarity of metal uptake routes within the taxonomic group of oligochaetes, either directly or indirectly, via the porewater compartment, only with different species-specific coefficients. However, the data for Cr, Cu, Ni, and Zn show a low impact of soil characteristics on metal uptake in *E. andrei* compared to *Enc. crypticus*, suggesting a lower influence of soil characteristics for the former organism. The most striking difference was found for Zn: Whereas internal Zn concentrations in *E. andrei* varied by no more than a factor of 5 (suggesting homeostasis), Zn concentrations in *Enc. crypticus* varied by as much as a factor of 27. Zinc uptake by *Enc. crypticus* could be described on the basis of a one-compartment model. Because the same soils were used under similar experimental conditions, this suggests that *E. andrei* has an increased capability over *Enc. crypticus* of regulating its internal Zn concentrations.

Table 5-3 Multivariate regression formulas relating (log-transformed) uptake parameters for Cd for *E. andrei* after exposure in Dutch soils to (log-transformed) external concentrations or soil and porewater parameters. (Significant characteristics are arranged in decreasing order of importance.)[a]

Normalization	Equations and range of validity of equations[b,c]	Statistics
Total soil concentration	$\log k_1(s) = $ —	no significant correlation
	$\log B_sAF = 3.15 - 0.38 \cdot pH - 0.56 \cdot \log DOC$	$R^2_{adj} = 0.61$, $n = 8$, s.e. = 0.19, $F = 6.4$, $P = 0.042$
Porewater concentration	$\log k_1(pw) = -0.59 + 0.49 \cdot pH$	$R^2_{adj} = 0.71$, $n = 12$, s.e. = 0.34, $F = 28.1$, $P < 0.001$
	$\log B_{pw}AF = $ —	no significant correlation
$CaCl_2$ extraction	$\log k_1(CaCl_2) = -3.19 + 0.65 \cdot pH$	$R^2_{adj} = 0.59$, $n = 12$, s.e. = 0.58, $F = 16.8$, $P = 0.002$
	$\log B_{CaCl_2}AF = 0.52 + 0.30 \cdot pH + 0.90 \cdot \log DOC$	$R^2_{adj} = 0.61$, $n = 8$, s.e. = 0.24, $F = 6.5$, $P = 0.041$
Porewater activity	$\log k_1(Cd^{2+}) = -1.02 + 0.63 \cdot pH$	$R^2_{adj} = 0.74$, $n = 12$, s.e. = 0.41, $F = 31.9$, $P < 0.001$
	$\log B_{Cd^{2+}}AF = $ —	no significant correlation
—	$\log C_w(ss) = 2.69 + 1.14 \cdot \log[Cd^{2+}]_{CaCl_2}$	$R^2_{adj} = 0.77$, $n = 8$, s.e. = 0.23, $F = 25.4$, $P = 0.002$
—	$\log C_w(ss) = -0.15 - 0.52 \cdot \log[Ca^{2+}]_{pw} + 0.45 \cdot \log[Cd^{2+}]_{CaCl_2}$	$R^2_{adj} = 0.92$, $n = 8$, s.e. = 0.14, $F = 41.9$, $P = 0.001$

[a] Data from Peijnenberg, Posthuma et al. 1999.

[b] Formula valid only for soils with $[Cd]_s > 0.01$ mmol/kg; otherwise, no significant uptake.

[c] Bioaccumulation factors (B_xAF) are relative to total soil concentrations, pore water (pw), $CaCl_2$ extraction ($CaCl_2$), or porewater Cd activity (Cd^{2+}). The uptake rate constant is k_1, ss indicates steady state, pH is pH($CaCl_2$), DOC is mmol/l.

Table 5-4 Multivariate regression formula relating (log-transformed) uptake parameters for Cd for *Enc. crypticus* after exposure in Dutch soils to (log-transformed) external concentrations or soil and porewater parameters. (Significant characteristics are arranged in decreasing order of importance.)[a]

Normalization	Equations and range of validity of equations[b,c]	Statistics
Total soil concentration	$\log k_1(s) = 1.36 - 0.35 \cdot pH$	$R^2_{adj} = 0.68$, $n = 19$, $F = 39.6$, $P < 0.001$, $Q^2 = 0.55$
	$\log k_1(s) = 1.81 - 0.42 \cdot pH$	$R^2_{adj} = 0.80$, $n = 18$, $F = 68.8$, $P < 0.001$, $Q^2 = 0.74$
	$\log B_sAF = 1.17 - 0.92 \cdot \log clay$	$R^2_{adj} = 0.64$, $n = 8$, $F = 13.2$, $P < 0.011$, $Q^2 = 0.59$
Porewater concentration	$\log k_1(pw) = —$	no significant correlation
	$\log B_{pw}AF = 3.83 + 0.25 \cdot pH$	$R^2_{adj} = 0.78$, $n = 9$, $F = 28.7$, $P < 0.001$, $Q^2 = 0.64$
$CaCl_2$ extraction	$\log k_1(CaCl_2) = —$	no significant correlation
	$\log B_{CaCl_2}AF = -5.76 + 2.46 \cdot \log OM$	$R^2_{adj} = 0.40$, $n = 9$, $F = 6.4$, $P < 0.040$, $Q^2 = 0.16$
Porewater activity	$\log k_1(Cd^{2+}) = —$	no significant correlation
	$\log B_{Me^{2+}}AF = 1.79 + 0.28 \cdot pH$	$R^2_{adj} = 0.88$, $n = 9$, $F = 61.3$, $P < 0.001$, $Q^2 = 0.83$
—	$\log C_w(ss) = 2.69 + 1.14 \cdot \log [Cd^{2+}]_{pw} + 0.27 \cdot pH$	$R^2_{adj} = 0.93$, $n = 9$, $F = 53.5$, $P < 0.001$, $Q^2 = 0.81$
—	$\log k_2 = —$	no significant correlation

[a] Source: Peijnenberg, Baerselman et al. 1999.

[b] Formula valid only for soils with $[Cd]_s > 0.01$ mmol/kg; otherwise, no significant uptake.

[c] Bioaccumulation factors (B_xAF) are relative to total soil concentrations, pore water (pw), $CaCl_2$ extraction ($CaCl_2$), or porewater Cd activity (Cd^{2+}). The uptake rate constant is k_1, ss indicates steady state, pH is pH($CaCl_2$), clay is clay content %, OM is organic matter %.

Summarizing, it was concluded by Peijnenburg, Baerselman et al. (1999) and Peijnenburg, Posthuma et al. (1999) that, although the focus was on porewater-mediated uptake, no conclusive evidence in support of this exposure route was found.

Outlook

Because different organisms have different uptake mechanisms and explore their heterogeneous habitats in different ways, the composite concept of "bioavailability" needs specification, both with respect to the characteristics and homogeneity of the soil and to the species (and ecotypes) involved. Species-specific aspects play a role in the uptake processes themselves, but they also factor in the redistribution processes inside the organism tissues. Further work should substantiate systematic relationships between uptake and soil characteristics for organisms other than the oligochaetes studied up to now. A variety of organisms should be studied, also focusing on those organisms for which uptake via the pore water is not obvious. In daily practice of risk assessment, grouping of species with similar uptake profiles seems most practical in the future, and an "averaged formula" for such groups should be developed to account for the differences among groups. To obtain a feasible approach, it can be foreseen that the species living on or in the soil need to be classified into groups with respect to their (complex) exposure routes. In addition, some insight should be obtained for the degree in which organisms change their environment, thereby altering the metal distribution in soil and inside the organism. For each group, a pragmatic formula can be derived, and site-specific exposure assessment then boils down to a characterization of the relative dominance of the different species. This may lead to the development of validated soil quality standards for which the issue of bioavailability has been specifically addressed. These standards will thus be predictive of adverse effects, depending on the relative frequencies of exposure routes on the one hand, and on species-specific reallocation and intoxication processes on the other. After all, the general aim of environmental management practices is not to protect a single worm species but to protect specific sites, specific processes, or life support functions within ecosystems. In addition, further attention should be paid to the issues of regulated versus nonregulated chemicals. True progress can be made only when the study of the chemical aspects of metal exposure is accompanied by the validation of the abiotic findings with toxicokinetic analyses for different organisms. Close cooperation between soil chemists and soil biologists is essential in this respect.

Chapter 6

Recommendations for Regulatory Programs and Research

Herbert E. Allen, Stephen P. McGrath, Michael J. McLaughlin,
Willie J.G.M. Peijnenburg, Sébastien Sauvé

Regulatory Programs

At present, our understanding of the factors that regulate the bioavailability of metals to organisms in soil is incomplete. We are unable to predict with relative certainty the impact that a specific concentration of metal will have on a soil of a specified physical and chemical composition. Nevertheless, we recognize that the effect of the metal depends on the physical and chemical properties of the soil. Consequently, we feel that regulatory programs should consider bioavailability, and as more definitive measures of bioavailability become available, these should be incorporated into soil screening levels, criteria, and standards.

As a result of the preparation of this report and in the discussions about bioavailability, the following 4 items were developed as guidance for regulatory programs:

1) Risk assessment should not be based on total metal concentrations alone. Bioavailability must be considered.
2) All pools of metal in soil do not have equal availability.
3) Bioavailability may change over time because of changes in soil conditions affecting metal availability. In addition, "aging" results in changes in metal distribution among pools and in bioavailability. Consequently, data from samples recently amended with soluble metal salts are unlikely to be predictive of effects of metal from field sites.
4) New methods for assessing metal bioavailability need to be incorporated more widely into regulatory frameworks, whether these are based on a fractional extraction of metal from soil (e.g., soil solution, saturation paste, 1.0 M NH_4NO_3) or on models that use total metal concentration and incorporate the major modifiers of toxicity. For this latter approach, studies should include and report measurements of total metal, extractable metal, soil solution pH, solid-phase metal oxide and organic matter content, dissolved organic matter (DOM), and Ca.

Bioavailability of Metals in Terrestrial Ecosystems: Importance of Partitioning for Bioavailability to Invertebrates, Microbes, and Plants.
Herbert E. Allen, editor. © 2002 Society of Environmental Toxicology and Chemistry (SETAC). ISBN 1-880611-46-5

Research Programs

A significant amount of research must be conducted to enable bioavailability to be incorporated into the regulatory process. We feel that the following 6 topics are the most important and should receive priority:

1) The relative importance of desorption of metal from the soil versus mass transfer through soil solution, versus diffusion to the biota, versus dissociation of metal complexes, versus uptake of metal by the organism should be established as a function of the important system parameters. This must be established in systems in which both soil and organisms are present, as opposed to hydroponic cultures, for example.
2) The Free Ion Activity Model (FIAM) for organisms in soil must be investigated to determine its predictive power.
3) Labile pools of metal for invertebrates should be determined.
4) Additional experimental data should be gathered for field-collected soils.
5) Methods should be developed to evaluate the pool of solid-phase metal that supplies metal to solution.
6) Methods should be developed that will allow the prediction of the risk that will be present in the future. This requires that changes in metal bioavailability resulting from changes in input to the environment and changes in modifying factors be incorporated. Changes in organism response at both the individual and community levels should be predicted.

Finally, we believe that close ties must be maintained between the scientific and regulatory communities to ensure relevancy and applicability. To advance the understanding of bioavailability, both chemistry and biology must be considered. If we do not consider both, it will not be possible to develop sufficiently robust predictions of bioavailability so that they can be used in regulatory programs.

Abbreviations

ACS	American Chemical Society
ASV	anodic stripping voltammetry
ATP	adenosine triphosphate
ATPase	adenosine triphosphatase
BAF	bioaccumulation factor
B_SAF	biota-to-soil accumulation factor
CEC	cation exchange capacity (Chapters 2 and 5), Commission of the European Communities (Chapter 4)
CCME	Canadian Council of Ministers of the Environment
CDTA	cyclohexyl-diamine-tetraacetate
CHEL	chelator
CSIRO	Commonwealth Scientific and Industrial Research Organization
DIN	Deutsch Institut für Normeng
DK-EPA	Danish Environmental Protection Agency
DOC	dissolved organic carbon
DOM	dissolved organic matter or material
DP-ASV	differential pulse-anodic stripping voltammetry
DTPA	diethylenetriaminepentaacetate
dw	dry weight
EC	effect concentration
EC*x*	effect concentration to *x*% of organisms
EDDA	ethylene-diamine-diacetate
EDTA	ethylenediaminetetraacetic acid
EGTA	ethylene-bis-(oxyethylenenitrilo)-tetraacetate
ETAP	Ecotoxicity Technical Advisory Panel

Bioavailability of Metals in Terrestrial Ecosystems: Importance of Partitioning for Bioavailability to Invertebrates, Microbes, and Plants.
Herbert E. Allen, editor. © 2002 Society of Environmental Toxicology and Chemistry (SETAC). ISBN 1–880611–46-5

EU	European Union
EXAFS	extended x-ray absorption fine structure
FIAM	Free Ion Activity Model
fw	fresh weight
HA	humic acid
HEDTA	hydroxyethyl-ethylenediamine-triacetate
HEIDA	hydroxyethyl-imino-diacetate
HEPES	4-(2-hydroxy-ethyl)-1-piperazineethanesulfonic acid
IACR	Institute of Arable Crops Research
ICA	International Copper Association
IEE	ion exchange equilibrium
ILZRO	International Lead Zinc Research Organization
ISE	ion-selective electrode
ISO	International Organization for Standardization
L_{ex}	excluded ligand
L_{inorg}	inorganic ligand
L_{org}	organic ligand
LC50	lethal concentration to 50% of organisms
LOAEC	lowest-observed-adverse-effect concentration
LP	lag period
M	molar
Me	metal
MMR	maximum mineralization rate
NAD^+	nicotinamide adenine dinucleotide
NADH	reduced form of NAD^+

NiPERA	Nickel Producers Environmental Research Association
NOAEC	no-observed-adverse-effect concentration
NTA	nitrilotriacetic acid
OECD	Organisation for Economic Cooperation and Development
OM	organic material
ORNL	Oak Ridge National Laboratory (USA)
RAC	variety of wheat, *Triticum aestivum*
S-EPA	Swedish Environmental Protection Agency
SETAC	Society of Environmental Toxicology and Chemistry
SOM	soil organic matter
UK	United Kingdom
USEPA	U.S. Environmental Protection Agency

References

Abd-Elfattah A, Wada K. 1981. Adsorption of lead, copper, zinc, cobalt, and cadmium by soils that differ in cation-exchange materials. *J Soil Sci* 32:271–283.

Abuzid MM, Obukhov AI. 1992. Effect of soil copper pollution on plants and the uptake of heavy metals by corn seedlings. *Mosc Univ Soil Sci Bull* 47:37–39.

Adriano DC. 1986. Trace elements in the terrestrial environment. New York NY: Springer-Verlag. 533 p.

Aldenberg T, Slob W. 1993. Confidence limits for hazardous concentrations based on logistically distributed NOEC toxicity data. *Ecotoxicol Environ Saf* 25:48–63.

Allen HE. 1993. The significance of metal speciation for water, sediment and soil quality standards. *Sci Tot Environ* (Suppl):23–45.

Allen HE, Boonlayangoor C, Noll KE. 1982. Changes in physicochemical forms of lead and calcium added to freshwater. *Environ Intern* 7:337–341.

Allen HE, Bell HE, Berry WJ, Di Toro DM, Hansen DJ, Meyer JS, Mitchell JF, Paquin PR, Reiley MC, Santore RS. 1999. Integrated approach to assessing the bioavailability and toxicity of metals in surface waters and sediments. Washington DC: USEPA, Office of Water and Office of Research and Development. EPA-822–E-99–001.

Allen HE, Hansen DJ. 1996. The importance of trace metal speciation to water quality criteria. *Water Environ Res* 68:42–54.

Allen HE, Lee S-Z, Huang CP, Sparks DL. 1994. The fate and transport of inorganic contaminants in New Jersey soils: Final report to New Jersey Department of Environmental Protection and Energy. Newark DE: Univ Delaware, Department of Civil and Environmental Engineering. 451 p.

Alloway BJ, Tills AR, Morgan H. 1984. The speciation and availability of cadmium and lead in polluted soils. *Trace Subst Environ Health* 18:187–201.

Amacher MC. 1984. Determination of ionic soil solutions and suspensions: Principal limitations. *Soil Sci Soc Am* 48:519–524.

Amann RI, Ludwig W, Schleifer KH. 1995. Phylogenetic identification and in situ detection of individual microbial cells without cultivation. *Microbiol Rev* 59:143–169.

Andersen C. 1979. Cadmium, lead, and calcium content, number and biomass in earthworms (*Lumbricidae*) from sewage sludge treated soil. *Pedobiologia* 19:309–319.

Anderson PR, Christensen TH. 1988. Distribution coefficients of Cd, Co, Ni, and Zn in soils. *J Soil Sci* 39:15–22.

Andersson A, Siman G. 1991. Levels of Cd and some other trace elements in soils and crops as influenced by lime and fertilizer level. *Acta Agric Scand* 41:3–11.

Angle JS, McGrath SP, Chaudri AM. 1992. Effects of media components on toxicity of Cd to rhizobia. *Water Air Soil Pollut* 64:627–633.

Apte SC, Natley GE. 1995. Trace metal speciation of labile chemical species in natural waters and sediments: Non-electrochemical approaches. In: Tessier A, Turner DR, editors. Metal speciation and bioavailability in aquatic systems. Chichester, UK: J Wiley. p 259-306.

Astruc M, Lecomte J, Mericam P. 1981. Evaluation of methods for speciation of heavy metals in water. *Environ Technol Lett* 2:1–8.

Atanassova I. 1995. Adsorption and desorption of Cu at high equilibrium concentrations by soil and clay samples from Bulgaria. *Environ Pollut* 87:17–21.

Aualiitia TU, Pickering WF. 1986. Anodic stripping voltametric study of the lability of Cd, Pb, Cu ions sorbed on humic acid particles. *Water Res* 20:1397–1406.

Avdeef A, Zabronsky J, Stuting H. 1983. Calibration of copper ion selective electrode response to pCu 19. *Anal Chem* 55:298–304.

Herbert E. Allen, editor. © 2002 Society of Environmental Toxicology and Chemistry (SETAC). ISBN 1–880611–46–5

Bååth E. 1989. Effects of heavy metals in soil on microbial processes and populations (a review). *Water Air Soil Pollut* 47:335–379.

Bååth E, Díaz-Raviña M, Frostegård A, Campbell CD. 1998. Effect of metal-rich sludge amendments on the microbial community. *Appl Environ Microbiol* 64:238–245.

Baes III CF, Sharp RD. 1983. A proposal for estimation of soil leaching and leaching constants for use in assessment models. *J Environ Qual* 12:17–28.

Balk F, Dogger JW, Noppert F, Rutten ALM, Hof M, Van Lamoen FBH. 1993. Methode voor de schatting van milieurisico's in de Gelderse uiterwaarden (in Dutch): Publications and reports of the project "Ecological rehabilitation of the rivers Rhine and Meuse;" nr 47. Lelystad, Netherlands: RIZA.

Barber I, Bembridge J, Dohmen P, Edwards P, Heimbach F, Heusel R, Romijn K, Rufli H. 1998. Development and evaluation of triggers for earthworm toxicity testing with plant protection products. In: Sheppard S, Bembridge J, Holmstrup M, Posthuma L, editors. Advances in earthworm ecotoxicology: Proceedings from the 2nd International Workshop on Earthworm Ecotoxicology; 2–5 April 1997; Amsterdam, Netherlands. Pensacola FL: SETAC. p 269–278.

Barber SA. 1984. Modeling nutrient uptake by plant roots growing in soil. In: Barber SA, editor. Soil nutrient bioavailability: A mechanistic approach. New York NY: Wiley-Interscience. p 114-135.

Barber SA. 1995. Soil nutrient bioavailability: A mechanistic approach. New York NY: J Wiley. 414 p.

Barnett MO, Harris LA, Turner RB, Henson TJ, Melton RE, Stevenson RJ. 1995. Characterization of mercury species in contaminated floodplain soils. *Water Air Soil Pollut* 80:1105–1108.

Barnett MO, Harris LA, Turner RB, Stevenson RJ, Henson TJ, Melton RE, Hoffman DF. 1997. Formation of mercuric sulfide in soil. *Environ Sci Technol* 31:3037–3043.

Barrow NJ, Bowden JW, Posner AM, Quirk JP. 1981. Describing the adsorption of copper, zinc, and lead on variable charge mineral surface. *Aust J Soil Res* 19:309–321.

Bartlett R, James B. 1980. Studying dried, stored soil samples: Some pitfalls. *Soil Sci Soc Am J* 44:721–724.

Basta NT, Pantone DJ, Tabatabai MA. 1993. Path analysis of heavy metal adsorption by soil. *Agron J* 85:1054–1057.

Bauer CF, Kheboian C. 1988. Response to comments on potential artifacts in the determination of metal partitioning in sediments by a sequential fractionation procedure. *Anal Chem* 60:1477.

Beckett PHT, Davis RD. 1977. Upper critical levels of toxic elements in plants. *New Phytol* 79:95-106.

Beckett PHT, Warr E, Davis RD. 1983. Cu and Zn in soils treated with sewage sludge: Their 'extractability' to reagents compared with their 'availability' to plants. *Plant Soil* 70:3–14.

Beckett PHT. 1989. The use of extractants in studies on trace metals in soils, sewage sludges, and sludge-treated soils. *Adv Soil Sci* 9:143–176.

Belfroid AC, Sijm DTHM, van Gestel CAM. 1995. Modeling the accumulation of hydrophobic organic chemicals in earthworms: Application of the equilibrium partitioning theory. *Environ Sci Pollut Res* 14:605–612.

Belfroid AC, Sijm DTHM, van Gestel CAM. 1996. Bioavailability and toxicokinetics of hydrophobic aromatic compounds in benthic and terrestrial invertebrates. *Environ Rev* 4:276-299.

Belfroid AC, van Gestel CAM. 1999. Blootstellingsroutes van toxische stoffen voor terrestrische invertebraten. Amsterdam, Netherlands: Free Univ. Report nr E-99–06. 51 p.

Bell PF, Chaney RL, Angle JS. 1991. Free metal activity and total metal concentration as indices of micronutrient availability to barley (*Hordeum vulgare* L. 'Klages'). *Plant Soil* 130:51–62.

Belli SL, Zirino A. 1993. Behavior and calibration of the copper-II ion-selective electrode in high chloride media and marine waters. *Anal Chem* 65:2583–2589.

Berggren D. 1989. Speciation of aluminum, cadmium, copper, and lead in humic soil solutions: A comparison of the ion exchange column procedure and equilibrium dialysis. *Intern J Environ Anal Chem* 35:1–15.

Bergkvist BO, Folkeson L, Berggren D. 1989. Fluxes of Cu, Zn, Pb, Cd, Cr, and Ni temperature forest ecosystems. *Water Air Soil Pollut* 47:217–286.

Bermond AP. 1992. Thermodynamics applied to the study of the limits of sequential extraction procedures used for the speciation of trace elements in sediments and soils. *Environ Technol* 13:1175–1179.

Bermond AP. 1993. The localization of heavy metals in soil samples: Thermodynamic or kinetic control? 9th International Conference Heavy Metals in the Environment. Volume 2; September 1993; Toronto, Canada. Edinburgh, UK: CEP Consultants Ltd. p 484–487.

Bernhard M, Brinckman FE, Sadler PJ. 1986. The importance of chemical speciation in environmental processes. Berlin, Germany: Springer-Verlag. 732 p.

Berti WR, Jacobs W. 1996. Chemistry and phytotoxicity of soil trace elements from repeated sewage sludge addition. *J Environ Qual* 25:1025–1032.

Beveridge A, Waller P, Pickering WF. 1989. Evaluation of "labile" metal in sediments using ion exchange resins. *Talanta* 36:535–542.

Bewley RJF, Stotzky G. 1983. Effects of cadmium and zinc on microbial activity in soil: Influence of clay minerals; Part I: Metals added individually. *Sci Tot Environ* 31:41–55.

Beyer WN, Chaney RL, Mulhern BM. 1982. Heavy metal concentrations in earthworms from soil amended with sewage sludge. *J Environ Qual* 11:381–385.

Beyer WN, Hensler G, Moore J. 1987. Relation of pH and other soil variables to concentrations of Pb, Cu, Zn, Cd, and Se in earthworms. *Pedobiologia* 30:167–172.

Bhat GA, Saar RA, Smart RB, Weber JH. 1981. Titration of soil-derived fulvic acid by copper(II) and measurement of free copper(II) by anodic stripping voltammetry and copper(II) selective electrode. *Anal Chem* 53:2274–2280.

Bierkens J, Klein G, Corbisier P, Van Den Heuvel R, Verschaeve L, Weltens R, Schoeters G. 1998. Comparative sensitivity of 20 bioassays for soil quality. *Chemosphere* 37:2935–2947.

Bierman PM, Rosen CJ, Nater EA. 1995. Soil solution chemistry of sewage-sludge incinerator ash and phosphate fertilizer amended soil. *J Environ Qual* 24:279–85.

Biggins PDE, Harrison RM. 1980. Chemical speciation of lead compounds in street dusts. *Environ Sci Technol* 14:336–339.

Bingham FT, Sposito G, Strong JE. 1984. The effect of chloride on the availability of cadmium. *J Environ Qual* 13:71–74.

Bingham FT, Sposito G, Strong JE. 1986. The effect of sulfate on the availability of cadmium. *Soil Sci* 141:172–177.

Bingham FT, Strong JE, Sposito G. 1983. Influence of chloride salinity on cadmium uptake by swiss chard. *Soil Sci* 135:160–165.

Binstock DA, Grohse PM, Gaskill Jr A, Sellers C, Luk KK, Kingston HM, Jassie LB. 1991. Development and validation of a method for determining elements in solid waste by microwave digestion. *Anal Chem* 74:360–366.

Blaedel WJ, Dinwiddie DE. 1974. Study of the behavior of copper ion-selective electrodes at submicromolar concentration levels. *Anal Chem* 46:873–879.

Blaylock MJ, Salt DE, Dushenkov S, Zakharova O, Gussman C, Kapulnik Y, Ensley BD, Raskin I. 1997. Enhanced accumulation of Pb in Indian mustard by soil-applied chelating agents. *Environ Sci Technol* 31:860–865.

Bouldin DR. 1989. A multiple ion uptake model. *J Soil Sci* 40:309–319.

Bowen JE. 1969. Absorption of copper, zinc, and manganese by sugarcane leaf tissue. *Plant Physiol* 44:55–261.

Bowers N, Pratt JR, Beeson D, Lewis M. 1997. Comparative evaluation of soil toxicity using lettuce seeds and soil ciliates. *Environ Toxicol Chem* 16:207–213.

Bresnahan WT, Grant CL, Weber JH. 1978. Stability constants for the complexation of copper(II) ions with water and soil fulvic acids measured by an ion selective electrode. *Anal Chem* 50:1675–1679.

Brewster JL, Tinker PB. 1970. Nutrient cation flows in soil around plant roots. *Soil Sci Soc Am Proc* 34:421–426.

Brookes PC, McGrath SP. 1984. The effects of metal toxicity on the size of the soil microbial biomass. *J Soil Sci* 35:341–346.

Brümmer GW. 1986. Heavy metal species, mobility, and availability in soils. In: Bernhard M, Brinckman FE, Sadler PJ, editors. The importance of chemical "speciation" in environmental processes. Berlin, Germany: Springer-Verlag. p 169–192.

Brümmer GW, Gerth J, Herms U. 1986. Heavy metals species, mobility and availability in soils. *Z Pflanzenernaehr Bodenkd* 149:382–398.

Brümmer GW, Gerth J, Tiller KG. 1988. Reaction kinetics of the adsorption and desorption of nickel, zinc and cadmium by goethite (I) Adsorption and diffusion of metals. *J Soil Sci* 39:37-52.

Brun LA, Maillet J, Richarte J, Herrman P, Remy JC. 1998. Relationships between extractable copper, soil properties and copper uptake in wild plants in vineyard soils. *Environ Pollut* 102:151–161.

Buchter B, Davidoff B, Amacher MC, Hinz C, Iskandar IK, Selim HM. 1989. Correlation of Freudlich *Kd* and *n* retention parameters with soils and elements. *Soil Sci* 148:370–379.

Buffle J. 1980. A critical comparison of studies of complex formation between copper(II) and fulvic substances of natural waters. *Anal Chim Acta* 118:29–44.

Buffle J, Greter FL, Haerdi W. 1977. Measurement of complexation properties of humic and fulvic acids in natural waters with lead and copper ion-selective electrodes. *Anal Chem* 49:216–222.

Cabaniss SE, Shuman MS. 1986. Combined ion selective electrode and fluorescence quenching detection for copper-dissolved organic matter titrations. *Anal Chem* 58:398.

Cabaniss SE, Shuman MS, Collins BJ. 1984. Metal-organic binding: A comparison of models. In: Kremer CJM, Duinker JC, editors. Complexation of trace metals in natural waters. The Hague, Netherlands: Martinus Nijhoff/W. Junk. p 165–179.

Cabrera D, Young SD, Rowell DL. 1988. The toxicity of cadmium to barley plants as affected by complex formation with humic acid. *Plant Soil* 105:195–204.

Camerlynck R, Kiekens L. 1982. Speciation of heavy metals in soils based on charge separation. *Plant Soil* 68:331–339.

Camobreco VJ, Richards BK, Steenhuis TS, Peverly JH, McBride M. 1996. Movement of heavy metals through undisturbed and homogenized soil columns. *Soil Sci* 161:740–750.

Campbell CD, Warren A, Cameron CM, Hope SJ. 1997. Direct toxicity assessment of two soils amended with sewage sludge contaminated with heavy metals using a protozoan (*Colpoda steinii*) bioassay. *Chemosphere* 34:501–514.

Campbell DJ, Beckett PHT. 1988. The soil solution in a soil treated with digested sewage sludge. *J Soil Sci* 39:283–298.

Campbell PGC. 1995. Interactions between trace elements and aquatic organisms: A critique of the free-ion activity model. In: Tessier A, Turner DR, editors. Metal speciation and bioavailability in aquatic systems. New York NY: J Wiley. p 45–102.

Camusso M, Tartari G, Zirino A. 1991. Measurement and prediction of copper ion activity in Lake Orta, Italy. *Environ Sci Technol* 25:678–683.

Cancio I, Gwynn I, Ireland MP, Cajaraville MP. 1995. The effect of sublethal lead-exposure on the ultrastructure and on the distribution of acid-phosphate-activity in chloragocytes of earthworms *Oligochaeta*. *Histochem J* 27:965–973.

Capelo S, Mota AM,Gonçalves ML. 1995. Complexation of lead with humic matter by anodic stripping voltammetry: Prevention of adsorption on nafion-coated mercury film electrode. *Electroanalysis* 7:563–568.

Carroll SA, O'Day PA, Piechowsky M. 1998. Rock-water interactions controlling zinc, cadmium, and lead concentrations in surface waters and sediments, U.S. Tri-State mining district. (2) Geochemical interpretation. *Environ Sci Technol* 32:956–965.

Carter A. 1983. Cadmium, copper, and zinc in soil animals and their food in a red clover system. *Can J Zool* 61:2751–2757.

Cataldo DA, Garland TR, Wildung RE. 1983. Cadmium uptake kinetics in intact soybean plants. *Plant Physiol* 73:844–848.

Cavallaro N, McBride MB. 1980. Response of the Cu^{2+} and Cd^{2+} ion-selective electrodes to solutions of different ionic strength and ion composition. *Soil Sci Soc Am J* 44:881–882.

[CCME] Canadian Council of Ministers of the Environment. Subcommittee on environmental quality criteria for contaminated sites. 1991. Interim Canadian environmental quality criteria for contaminated sites. Winnipeg Manitoba, Canada: CCME. EPC-CS34. 107 p.

[CEC] Commission of the European Communities. 1986. On the protection of the environment, and in particular of the soil, when sewage sludge is used in agriculture. *Off J Eur Communities*. nr L81: Annex 1A:6–12.

Chaney RL. 1980. Health risks associated with toxic metals in municipal sludge. In: Bitton G, Damro DL, Davidson GT, Davidson JM, editors. Sludge: Health risks of land application. Ann Arbor MI: Ann Arbor Science. p 59–83.

Chang AC, Granato TC, Page AL. 1992. A methodology for establishing phytotoxicity criteria for chromium, copper, nickel, and zinc in agricultural land application of municipal sewage sludges. *J Environ Qual* 21:521–536.

Chang AC, Page AL, Warneke JEGE. 1984. A sequential extraction of soil heavy metals following a sludge application. *J Environ Qual* 13:33–38.

Chao T. 1972. Selective dissolution of manganese oxides from soils and sediments with acidified hydroxylamine hydrochloride. *Soil Sci Soc Am Proc* 36:764–768.

Chapman PM, Caldwell RS, Chapman PF. 1996. A warning: NOECs are inappropriate for regulatory use. *Environ Toxicol Chem* 15:77–79.

Chaudhri MB. 1993. Determination and calculation of metal ion activities from metal ion-chelate equilibria. *Sarhad J Agric* 9:247–252.

Chaudhry FM, Longeragan JF. 1972a. Zinc absorption by wheat seedlings and the nature of its inhibition by alkaline earth cations. *J Experiment Bot* 23:552–560.

Chaudhry FM, Longeragan JF. 1972b. Zinc absorption by wheat seedlings: (I) Inhibition by macronutrient ions in short-term experiments and its relevance to long-term zinc nutrition. *Soil Sci Soc Am Proc* 36:323–331.

Chaudhry FM, Longeragan JF. 1972c. Zinc absorption by wheat seedlings: (II) Inhibition by hydrogen ions and by micronutrient cations. *Soil Sci Soc Am Proc* 36:327–331.

Chaudri AM, Knight BP, Barbosa-Jefferson VL, Preston S, Paton GI, Killham K, Coad N, Nicholson FA, Chambers BJ, McGrath SP. 1999. Determination of acute Zn toxicity in pore water from soils previously treated with sewage sludge using bioluminescence assays. *Environ Sci Technol* 33:1880–1885.

Chaudri AM, McGrath SP, Giller KE. 1992. Survival of the indigenous population of *Rhizobium leguminosarum* biovar *trifolii* in soil spiked with Cd, Zn, Cu, and Ni salts. *Soil Biol Biochem* 24:625–632.

Chaudri AM, McGrath SP, Giller KE, Rietz E, Sauerbeck DR. 1993. Enumeration of indigenous *Rhizobium leguminosarum* biovar *trifolii* in soils previously treated with metal-contaminated sewage sludge. *Soil Biol Biochem* 25:301–309.

Checkai RT, Corey RB, Helmke PA. 1987. Effects of ionic and complexed metal concentrations on plant uptake of cadmium and micronutrient metals from solution. *Plant Soil* 99:335-345.

Chen M, Ma LQ. 1998. Comparison of four USEPA digestion methods for trace metals analysis using certified and Florida soils. *J Environ Qual* 27:1294-1300.

Chen Z, Mayer LM. 1998. Mechanisms of Cu solubilization during deposit feeding. *Environ Sci Technol* 32:770–775.

Christensen BT. 1992. Physical fractionation of soil and organic matter in primary particle size and density separates. *Adv Soil Sci* 20:1–90.

Christensen TH. 1984. Cadmium soil sorption at low concentrations (I) Effect of time, cadmium load, pH, and calcium. *Water Air Soil Pollut* 21:105–114.

Christensen TH. 1987. Cadmium soil sorption at low concentrations (V) Evidence of competition by other heavy metals. *Water Air Soil Pollut* 34:293–303.

Clarkson DT. 1988. Movement of ions across roots. In: Baker DA, Hall JL, editors. Solute transport in plant cells and tissues. Marlow, UK: Longman. p 251–304.

Clarkson DT. 1993. Roots and the delivery of solutes to the xylem. *Phil Tran Royal Soc London* B341:5–17.

Colwell RR. 1979. Enumeration of specific populations by the most-probable number (MPN) method. In: Costerton JW, Colwell RR, editors. Native aquatic bacteria: Enumeration, activity and ecology. Philadelphia PA: ASTM. STP 695. p 56–61.

Cook N. 1997. Bioavailability of trace metals in urban contaminated soil [Ph.D. Dissertation]. Montreal Canada: McGill Univ-Macdonald Campus. Department of Natural Resource Sciences. 128 p.

Cook N, Hendershot WH. 1996. The problem of establishing ecologically based soil quality criteria: The case of lead. *Can J Soil Sci* 76:335–342.

Cooper KM, Tinker PB. 1978. Translocation and transfer of nutrients in vesicular-arbuscular mycorrhizas: (II) Uptake and translocation of phosphorus, zinc, and sulfur. *New Phytol* 73:901-912.

Corp N, Morgan AJ. 1991. Accumulation of heavy metals from polluted soils by the earthworm *Lumbricus rubellus*: Can laboratory exposure of 'control' worms reduce biomonitoring problems? *Environ Pollut* 74:39–52.

Costa G, Morel JL. 1993. Cadmium uptake by *Lupinus albus* (L.): Cadmium excretion, a possible mechanism of cadmium tolerance. *J Plant Nutrition* 16:1921–1929.

Crommentuijn T, Doodeman CJAM, Van der Pol JJC, Doornekamp A, Rademaker MCJ, van Gestel CAM. 1994. Lethal body concentrations and accumulation patterns determine time-dependent toxicity of cadmium in soil arthropods. *Environ Toxicol Chem* 13:1781–1789.

Crommentuijn T, Doodeman CJAM, Van der Pol JJC, Doornekamp A, van Gestel CAM. 1997. Bioavailability and ecological effects of cadmium on *Folsomia candida* (Willem) in an artificial soil substrate as influenced by pH and organic matter. *Appl Soil Ecol* 5:261–271.

Crowdy SH, Tanton TW. 1970. Water pathways in higher plants. *J Experiment Bot* 21:102–111.

Dahlin S, Witter W, Mårtensson A, Turner A, Bååth E. 1997. Where's the limit? Changes in microbiological properties of agricultural soils at low levels of metal contamination. *Soil Biol Biochem* 29:405–1415.

Dallinger R. 1977. The flow of copper through a terrestrial food chain (III): Selection of an optimum copper diet by isopods. *Oecologia* 30:271–276.

Dallinger R. 1993. Strategies of metal detoxification in terrestrial invertebrates. In: Dallinger R, Rainbow PS, editors. Ecotoxicology of metals in invertebrates. Chelsea MI: Lewis. p 245–289.

Dallinger R, Wieser W. 1977. The flow of copper through a terrestrial food chain (I): Copper and nutrition in isopods. *Oecologia* 30:253–264.

Davenport JR, Peryea FJ. 1991. Phosphate fertilizers influence leaching of lead and arsenic in a soil contaminated with lead arsenate. *Water Air Soil Pollut* 57–58:101.

Davies BE. 1992. Inter-relationships between soil properties and the uptake of cadmium, copper, lead and zinc from contaminated soils by radish (*Raphanus sativus* L.). *Water Air Soil Pollut* 63:331–342.

Davis RD, Beckett PHT. 1978. Upper critical levels of toxic elements in plants. *New Phytol* 80:23-32.

de Groot AC, Peijnenburg WJGM, Van den Hoop MAGT, Van Veen RPM. 1998. Heavy metals in Dutch field soils: An experimental and theoretical study on equilibrium partitioning. Bithoven, Netherlands: RIVM. Report nr 607220 001. 46 p.

de Haan FAM, Van der Zee SEATM, Van Riemsdjik WH. 1987. The role of soil chemistry and soil physics in protecting soil quality: Variability of sorption and transport of cadmium as an example. *Nether J Agric Sci* 35:87.

Dean RB, Suess MJ. 1985. The risk to health of chemicals in sewage sludge applied to land. *Waste Manag Res* 3:251–278.

Deaver E, Rodgers Jr JH. 1996. Measuring bioavailable copper using anodic stripping voltammetry. *Environ Toxicol Chem* 15:1925–1930.

DeKock PC. 1956. Heavy metal toxicity and iron chlorosis. *Annals Bot* 20:133–141.

DeKock PC, Mitchell RL. 1957. Uptake of chelated metals by plants. *Soil Sci* 84:55–63.

Delhaize E, Ryan PR, Randall PJ. 1993. Aluminum tolerance in wheat (*Triticum aestivum* L.). (II) Aluminum-stimulated excretion of malic acid from root apices. *Plant Physiol* 103:695–702.

Descamps M, Fabre MC, Grelle C, Gerard S. 1996. Cadmium and lead kinetics during experimental contamination and decontamination of the centipede *Lithobius forficatus* L. *Arch Environ Contam Toxicol* 31:350–353.

Díaz-Raviña M, Bååth E. 1996a. Development of metal tolerance in soil bacterial communities exposed to experimentally increased metal levels. *Appl Environ Microbiol* 62:2970–2977.

Díaz-Raviña M, Bååth E. 1996b. Thymidine and leucine incorporation into bacteria from soils experimentally contaminated with heavy metals. *Appl Soil Ecol* 3:215–224.

Díaz-Raviña M, Bååth E, Frostegård A. 1994. Multiple heavy metal tolerance of soil bacterial communities and its measurement by a thymidine incorporation technique. *Appl Environ Microbiol* 60:2238–2247.

Dietze G, König N. 1988. Metallspeziierung in bodenlösungen mittels dialysen- und ionenaustauscherverfahren. *Z Pflanzenernaehr Bodenkd* 151:243–250.

[DIN] Deutsch Institut fur Normeng. 1995. Soil quality extraction of trace elements with ammoniumnitrate solution. Berlin, Germany: DIN. DIN 19730.

Di Toro DM, Allen HE, Bergman HL, Meyer JS, Paquin PR, Santore RC. 2001. Biotic ligand model of the acute toxicity of metals. I. Technical basis. *Environ Toxicol Chem* 20:2383–2396.

Doelman P, Haanstra L. 1984. Short-term and long-term effects of cadmium, chromium, copper, nickel, lead, and zinc on soil microbial respiration in relation to abiotic soil factors. *Plant Soil* 79:317–327.

Doner HE. 1978. Chloride as a factor in mobilities of Ni(II), Cu(II), and Cd(II) in soil. *Soil Sci Soc Am J* 42:882–885.

Dowdy DL, McKone TE. 1997. Predicting plant uptake of organic chemicals from soil or air using octanol/water and octanol/air partition ratios and a molecular connectivity index. *Environ Toxicol Chem* 16:2448–2456.

Ducaroir J, Cambier P, Leydecker J, Prost R. 1990. Application of soil fractionation methods to the study of the distribution of pollutants metals. *Z Pflanzenernaehr Bodenkd* 153:349–358.

Ducaroir J, Lamy I. 1995. Evidence of trace metal association with soil organic matter using particle size fractionation after physical dispersion treatment. *Analyst* 120:741–745.

Dueck TA, Visser P, Ernst WHO, Schat H. 1986. Vesicular-arbuscular mycorrhizae decrease zinc-toxicity to grasses growing in zinc-polluted soil. *Soil Biol Biochem* 18:331–333.

Dumestre A, Sauvé S, McBride M, Bavege P, Berthelin J. 1999. Copper speciation and microbial activity in long-term contaminated soils. *Arch Environ Contam Toxicol* 36:124–131.

Dunemann L, von Wiren N, Schulz R, Marschner H. 1991. Speciation analysis of nickel in soil solutions and availability to oat plants. *Plant Soil* 133:263–270.

Duquette M, Hendershot WH. 1990. Copper and zinc sorption on some B horizons of Quebec soils. *Commun Soil Sci Plant Anal* 21:377–394.

Dzombak DA, Fish W, Morel FMM. 1986. Metal-humate interactions. (1) Discrete ligand and continuous distribution models. *Environ Sci Technol* 20:669–675.

Echeverría JC, Morera MT, Mazkiarán C, Garrido JJ. 1998. Competitive sorption of heavy metals by soils. Isotherms and fractional factorials experiments. *Environ Pollut* 101:275–284.

Edwards CA, Lofty JR. 1977. Biology of earthworms. 2nd ed. London, England: Chapman and Hall. 333 p.

El-Falaky AA, Aboulroos SA, Lindsay WL. 1991. Measurement of cadmium activities in slightly acidic to alkaline soils. *Soil Sci Soc Am J* 55:974–979.

Elkhatib EA, Bennett OL, Baligar VC, Wright RJ. 1986. A centrifuge method for obtaining soil solution using an immiscible liquid. *Soil Sci Soc Am J* 50:297–299.

Elliott HA, Shields GA. 1988. Comparative evaluation of residual and total metal analyses in polluted soils. *Commun Soil Sci Plant Anal* 19:1907–1916.

Elsokkary IH. 1980. Selenium distribution, chemical fractionation and adsorption in some egyptian alluvial and lacustrine soils. *Z Pflanzenernaehr Bodenkd* 143:74–83.

Elzinga EJ, Van Grinsven JJM, Swartjes FA. 1999. General purpose Freundlich isotherms for cadmium, copper and zinc in soils. *Eur J Soil Sci* 50:139–149.

Ephraim J, Alegret S, Mathuthu A, Bicking M, Malcolm RL, Marinsky JA. 1986. A unified physicochemical description of the protonation and metal ion complexation equilibria of natural organic acids (humic and fulvic acids). (2) Influence of polyelectrolyte properties and functional group heterogeneity on the protonation equilibria of fulvic acid. *Environ Sci Technol* 20:354–366.

Ephraim J, Marinsky JA. 1986a. A unified physico-chemical description of the protonation and metal ion complexation equilibria of natural organic acids (humic and fulvic acids). (1) Analysis of the influence of polyelectrolyte properties on protonation equilibria in ionic media: Fundamental concepts. *Environ Sci Technol* 20:349–366.

Ephraim J, Marinsky JA. 1986b. A unified physicochemical description of the protonation and metal ion complexation equilibria of natural organic acids (humic and fulvic acids). (3) Influence of polyelectrolyte properties and functional heterogeneity on the copper ion binding in an Armadale horizons Bh fulvic acid sample. *Environ Sci Technol* 20:367–376.

Escrig I, Morell I. 1998. Effect of calcium on the soil adsorption of cadmium and zinc in some Spanish sandy soils. *Water Air Soil Pollut* 105:507–520.

Esnaola MV, Millán E. 1998. Evaluation of heavy metal lability in polluted soils by a cation exchange batch procedure. *Environ Pollut* 99:79–86.

Essington ME, Mattigod SV. 1990. Element partitioning in size- and density-fractionated sewage sludge and sludge-amended soil. *Soil Sci Soc Am J* 54:385–394.

Essington ME, Mattigod SV. 1991. Trace element solid-phase associations in sewage sludge and sludge-amended soil. *Soil Sci Soc Am J* 55:350–356.

Evans LJ. 1989. Chemistry of metal retention by soils. *Environ Sci Technol* 23:1046–1056.

Evans LJ, Spiers GA, Zhao G. 1995. Chemical aspects of heavy metal solubility with reference to sewage sludge amended soils. *Intern J Environ Anal Chem* 59:291–302.

Figura P, McDuffie B. 1979. Use of chelex resin for determination of labile trace metal fractions in aqueous ligand media and comparison of the method with anodic stripping voltammetry. *Anal Chem* 51:120–125.

Fish W, Dzombak DA, Morel FMM. 1986. Metal-humate interactions. (2) Application and comparison of models. *Environ Sci Technol* 20:676–683.

Fließbach A, Reber H. 1991. Auswirkungen einen langjahrigan zufur von klarschlamm auf bodermikroorganismen und ihre leistungen? In: Sauerbeck DR, Lübben S, editors. Auswirkungen von siedlungsabfällen auf böden, bodenorganismen und pflanzen. Forschungszentrum Jülich GmbH. Berichte aus der Ökologischen Forschung. p 327–358.

Florence TM. 1982. The speciation of trace elements in waters. *Talanta* 29:345–364.

Florence TM. 1986. Electrochemical approaches to trace element speciation in waters: A review. *Analyst* 111:489–505.

Florence TM. 1989. Electrochemical techniques for trace element speciation in waters. In: Batley GE, editor. Trace element speciation: Analytical methods and problems. Boca Raton FL: CRC Pr. p 77–116.

Florence TM, Batley GE. 1980. Chemical speciation in natural waters. *CRC Crit Rev Anal Chem* 9:219–296.

Flores-Vélez LM, Ducaroir J, Jaunet AM, Robert M. 1996. Study of the distribution of copper in an acid sandy vineyard soil by three different methods. *Eur J Soil Sci* 47:523–532.

Förstner U. 1995. Land contamination by metals: Global scope and magnitude of problem. In: Allen HE, Huang CP, Bailey GW, Bower AR, editors. Metal speciation and contamination of soil. Boca Raton FL: Lewis. p 1–33.

Fortin C, Campbell PGC. 1998. An ion-exchange technique for free-metal ion measurements (Cd^{2+}, Zn^{2+}): Applications to complex aqueous media. *Intern J Environ Anal Chem* 72:173-194.

Fotovat A, Naidu R. 1997. Ion exchange resin and MINTEQA2 speciation of Zn and Cu in alkaline sodic and acidic soils extracts. *Aust J Soil Res* 35:711–726.

Fotovat A, Naidu R. 1998. Changes in composition of soil aqueous phase influence chemistry of indigenous heavy metals in alkaline sodic and acidic soils. *Geoderma* 84:213–234.

Frost RR, Griffin RA. 1977. Effect of pH on adsorption of arsenic and selenium from landfill leachate by clay minerals. *Soil Sci Soc Am J* 41:53–57.

Fruchter JS, Rai D, Zachara JM. 1990. Identification of solubility-controlling solid phases in a large fly ash field lysimeter. *Environ Sci Technol* 24:1173–1179.

Fujii R, Hendrickson LL, Corey RB. 1983. Ionic activities of trace metals in sludge-amended soils. *Sci Tot Environ* 28:179–190.

Fulghum JE, Bryan SR, Linton RW, Bauer CF, Griffis DP. 1988. Discrimination between adsorption and coprecipitation in aquatic particle standards by surface analysis techniques: Lead distribution in calcium carbonates. *Environ Sci Technol* 22:463–467.

Galli U, Schuepp H, Brunold C. 1994. Heavy metal binding by mycorrhizal fungi. *Physiologia Plantarum* 92:364–368.

Gamble DS, Langford H, Undertown AW. 1984. The interrelationship of aggregation and cation binding of fulvic acid. In: Kramer CJM, Duinker JC, editors. Complexation of trace metals in natural waters. The Hague, Netherlands: Martinus Nijhoff/W. Junk. p 349–356.

Gamble DS, Underdown AW, Langford CH. 1980. Copper (II) titration of fulvic acid ligand sites with theoretical, potentiometric and spectrophotometric analysis. *Anal Chem* 52:1901–1908.

Gambrell M. 1993. Cadmium accumulation in *Folsomia candida* [Collembola] and its effects upon growth and reproduction [Student's report]. Amsterdam, Netherlands: Free Unversity.

Ge Y, Murray P, Hendershot WH. 2000. Trace metal speciation and bioavailability in urban soils. *Environ Pollut* 107:137–144.

Gerritse RG, Van Driel W. 1984. The relationship between adsorption of trace metals, organic matter, and pH in temperate soils. *J Environ Qual* 13:197–204.

Gerritse RG, Van Driel W, Smilde KW, Van Luit B. 1983. Uptake of heavy metals by crops in relation to their concentration in the soil solution. *Plant Soil* 75:393–404.

Giller KE, McGrath SP, Hirsch PR. 1989. Absence of nitrogen fixation in clover grown on soil subject to long term contamination with heavy metals is due to survival of only ineffective *Rhizobium. Soil Biol Biochem* 21:841–848.

Giller KE, Witter E, McGrath SP. 1998. Toxicity of heavy metals to microorganisms and microbial processes in agricultural soils: A review. *Soil Biol Biochem* 30:1389–1414.

Giordano PM, Mays DA, Behel AD. 1979. Soil temperature effects on uptake of cadmium and zinc by vegetables grown on sludge-amended soil. *J Environ Qual* 8:233–236.

Giordano PM, Noggle JC, Mortvedt JJ. 1974. Zinc uptake by rice, as affected by metabolic inhibitors and competing cations. *Plant Soil* 41:637–646.

Gonçalves ML, Sigg L, Reutlinger M, Stumm W. 1987. Metal ion binding by biological surfaces voltammetric assessment in the presence of bacteria. *Sci Tot Environ* 60:105–120.

Gonçalves MLS, Sigg L, Stumm W. 1985. Voltammetric methods for distinguishing between dissolved and particulate metal ion concentrations in the presence of hydrous oxides. A case study on lead(II). *Environ Sci Technol* 19:141–146.

Gooddy DC, Shand P, Kinniburgh DG, Van Riemdsjik WH. 1995. Field-based partition coefficients for trace elements in soil solutions. *Eur J Soil Sci* 46:265–285.

Gries D, Brunn S, Crowley DE, Parker DR. 1995. Phytosiderophore release in relation to micronutrient metal deficiencies in barley. *Plant Soil* 172:299-308.

Gulens J. 1987. Assessment of research on the preparation, response, and application of solid-state copper ion-selective electrodes. *Ion-Sel Electrode Rev* 9:127–171.

Gulens J, Leeson PK, Séguin L. 1984. Kinetics influences on studies of copper (II). Hydrolysis by copper ion-selective electrode. *Anal Chim Acta* 156:19–31.

Gunn AM, Winnard DA, Hunt DTE. 1988. Trace metal speciation in sediments and soils. In: Kramer JR, Allen HE, editors. Metal speciation: Theory, analysis, and application. Chelsea MI: Lewis. p 261–294.

Gupta GC, Harrison FL. 1981. Effect of cations on copper adsorption by kaolin. *Water Air Soil Pollut* 15:323–327.

Gupta SK. 1984. Importance of soil solution composition in deciding the best suitable analytical criteria for guidelines on maximum tolerable metal load and in assessing bio-significance of metals in soil. *Schweiz Landw Fo* 23:209–225.

Gupta SK, Aten C. 1993. Comparison and evaluation of extraction media and their suitability in a simple model to predict the biological relevance of heavy metal concentrations in contaminated soils. *Int J Environ Anal Chem* 51:25–46.

Gutknecht J. 1981. Inorganic mercury (Hg^{2+}) transport through lipid bilayer membranes. *J Membrane Biol* 61:61–66.

Haddad KS, Evans JC. 1993. Assessment of chemical methods for extracting zinc, manganese, copper, and iron from New South Wales soils. *Commun Soil Sci Plant Anal* 24:29–44.

Haghiri F. 1974. Plant uptake of cadmium as influenced by cation exchange capacity, organic matter, zinc, and soil temperature. *J Environ Qual* 3:180–183.

Hale VQ, Wallace A. 1970. Effect of chelates on uptake of some heavy metal radionuclides from soil by bush beans. *Soil Sci* 109:262–263.

Halvorson AD, Lindsay WL. 1977. The critical Zn^{2+} concentration for corn and the nonabsorption of chelated zinc. *J Soil Sci Soc Am* 41:531–534.

Hamel SC, Buckley B, Lioy PJ. 1998. Bioaccessibility of metals in soils for different liquid to solid ratios in synthetic gastric fluid. *Environ Sci Technol* 32:358–362.

Hamelink JL, Landrum PF, Bergman HL, Benson WH. 1994. Bioavailability: Physical, chemical, and biological interactions. Boca Raton FL: CRC. 239 p.

Hamon RE, Lorenz SE, Holm TH, McGrath SP. 1995. Changes in trace metal species and other components of the rhizosphere during growth of radish. *Plant Cell Environ* 18:749-756.

Hamon RE, Wunke J, McLaughlin MJ, Naidu R. 1997. Availability of zinc and cadmium to different plant species. *Aust J Soil Res* 35:1267–1277.

Häni H, Gupta S. 1982. Total and biorelevant heavy metal contents and their usefulness in establishing limiting values in soils. In: Davis RD, Hucker G, L'Hermite P, editors. Environmental effects of organic and inorganic contaminants in sewage sludge. Dordrecht, Netherlands: D Reidel. p 121–129.

Häni H, Gupta S. 1985a. Chemical methods for the biological characterization of metal in sludge and soil. In: L'Hermite P, editor. Processing and use of organic sludge and liquid agricultural wastes. Dordrecht, Netherlands: D Reidel. p 157–167.

Häni H, Gupta S. 1985b. Reasons to use neutral salt solutions to assess the metal impact on plant and soils. In: Leschber R, Davis RD, L'Hermite P, editors. Chemical methods for assessing bioavailable metals in sludge and soils. London, England: Elsevier. p 42–47.

Hansen EH, Lamm CG, Rùzicka J. 1972. Selectrode - The universal ion-selective electrode. (II) Comparison of copper (II) electrodes in metal buffers and complexometric titrations. *Anal Chim Acta* 59:403–425.

Hansen EH, Rùzicka J. 1974. Selectrode -The universal ion-selective electrode Part VIII. The solid-state lead(II) selectrode in lead(II) buffers and potentiometric titrations. *Anal Chim Acta* 72:365–373.

Haq AU, Bates TE, Soon YK. 1980. Comparison of extractants for plant-available zinc, cadmium, nickel, and copper in contaminated soils. *Soil Sci Soc Am J* 44:772–777.

Harrison SJ, Lepp NW, Phipps DA. 1984. Uptake of copper by excised roots. (IV) Copper uptake from complexed sources. *Z Planzenphysiol* 113:445–450.

Hart BT. 1981. Trace metal complexing capacity in natural waters: A review. *Environ Technol Lett* 2:95–110.

Hart BT, Jones MJ. 1984. Measurement of the trace metal complexing capacity of Magela Creek waters. In: Kramer CJM, Duinker JC, editors. Complexation of trace metals in natural waters. The Hague, Netherlands: Martinus Nijhoff/W. Junk. p 201–211.

Hart JJ, Welch RM, Norvell WA, Sullivan LA, Kochian LV. 1998. Characterization of cadmium binding, uptake, and translocation in intact seedlings of bread and durum wheat cultivars. *Plant Physiol* 116:1413–1420.

Hartenstein R. 1964. Feeding, digestion, glycogen and the environmental conditions in the digestive system in *Oniscus asellus*. *J Insect Physiol* 10:611–621.

Harter RD 1991a. Kinetics of sorption/desorption processes in soil. In: Sparks DL, Suarez DL, editors. Kinetics of soil chemical processes. Madison WI: Soil Science Society of America. p 135–150.

Harter RD 1991b. Micronutrient adsorption-desorption reactions in soils. In: Mortvedt JJ, Cox FR, Shuman LM, Welch RM, editors. Micronutrients in agriculture. 2nd ed. SSSA Book Series #4. Madison WI: Soil Science Society of America. p 59–87.

Harter RD, Naidu R. 1995. Role of metal-organic complexation in metal sorption by soils. *Adv Agron* 55:219–263.

Hatch DJ, Jones LHP, Burau RG. 1988. The effect of pH on the uptake of cadmium by four plant species grown in flowing solution culture. *Plant Soil* 105:121–126.

Heggo A, Angle JS, Chaney RL. 1990. Effects of vesicular-arbuscular mycorrhizal fungi on heavy metal uptake by soybeans. *Soil Biol Biochem* 22:865–869.

Heikens A, Peijnenberg WJGM, Hendriks AJ. 2001. Bioaccumulation of heavy metals in terrestrial invertebrates. *Environ Pollut* 113:385–393.

Helmke PA, Robarge WP, Korotev R, Schomberg PJ. 1979. Effects of soil-applied sewage sludge on concentrations of elements in earthworms. *J Environ Qual* 8:322–327.

Hendershot WH, Courchesne F. 1991. Comparison of soil solution chemistry in zero tension and ceramic-cup tension lysimeters. *J Soil Sci* 42:577–584.

Hendrickson LL, Corey RB. 1983. A chelating-resin method for characterizing soluble metal complexes. *Soil Sci Soc Am J* 47:467–474.

Hering JG. 1995. Implications of complexation, sorption and dissolution kinetics for metal transport in soils. In: Allen HE, Huang CP, Bailey GW, Bowers AR, editors. Metal speciation and contamination of soil. Boca Raton FL: Lewis. p 59–86.

Hering JG, Morel FMM. 1988. Kinetics of trace metal complexation: Role of alkaline earth metals. *Environ Sci Technol* 22:1469–1478.

Hering JG, Morel FMM. 1989. Slow coordination reactions in seawater. *Geochim Cosmochim Acta* 53:611–618.

Hering JG, Morel FMM. 1990a. The kinetics of trace metal complexation: Implications for metal reactivity in natural waters. In: Stumm W, editor. Aquatic chemical kinetics: Reaction rates of processes in natural waters. New York NY: Wiley Interscience. p 145–171.

Hering JG, Morel FMM. 1990b. Kinetics of trace metal complexation: Ligand-exchange reactions. *Environ Sci Technol* 24:242–252.

Hollis L, Muench L, Playle RC. 1997. Influence of dissolved organic matter on copper binding, and calcium on cadmium binding, by gills of rainbow trout. *J Fish Biol* 50:703–720.

Holm PE, Christensen TH, Lorenz SE, Hamon RE, Domingues HC, Sequeira EM, McGrath SP. 1998. Measured soil water concentrations of cadmium and zinc in plant pots and estimated leaching outflows from contaminated soils. *Water Air Soil Pollut* 102:105–115.

Holm PE, Christensen TH, Tjell JC, McGrath SP. 1995. Heavy metals and the environment: Speciation of cadmium and zinc with application to soil solutions. *J Environ Qual* 24:183-190.

Holm TR. 1990. Copper complexation by natural organic matter in contaminated and uncontaminated ground water. *Chem Speciat Bioavailab* 2:63–76.

Holm TR, Curtiss III CD. 1990. Copper complexation by natural organic matter in ground water. In: Melchior DC, Bassett RL, editors. Chemical modeling of aqueous systems II. Washington DC: ACS. ACS symposium series 416. p 508–518.

Hopkin SP. 1989. Ecophysiology of metals in terrestrial invertebrates. London, England: Elsevier Applied Science. 366 p.

Houba VJG, Novozamsky I, Lexmond ThM, Van der Lee JJ. 1990. Applicability of 0.01 M $CaCl_2$ as a single extraction solution for the assessment of the nutrient status of soils and other diagnostic purposes. *Commun Soil Sci Plant Anal* 21:2281–2290.

Huang JW, Chen J, Berti WR, Cunningham SD. 1997. Phytoremediation of lead-contaminated soils: Role of synthetic chelates in lead phytoextraction. *Environ Sci Technol* 31:800–805.

Hubbell SP, Sikora A, Paris OH, 1965. Radiotracer, gravimetric, and calorimetric studies of ingestion and assimilation rates of an isopod. *Health Phys* 11:1485–1501.

Hughes MN, Poole RK. 1989. Metals and micro-organisms. New York NY: Chapman and Hall. 412 p.

Humbert W. 1974. Étude du pH intestinal d'un Collembole (insecte, apterygote). *Rev Ecol Biol Sol* 11:89–97.

Indeherberg MBM, De Vocht AJP, van Gestel CAM. 1998. Biological interactions: Effects on and the use of soil invertebrates in relation to soil contamination and in situ soil reclamation. In: Vangronsveld J, Cunningham SD, editors. Metal-contaminated soils: In situ inactivation and phytorestoration. Austin TX: Landes Bioscience. p 93–119.

[ISO] International Organization for Standardization. 1998. Soil quality-effects of pollutants on earthworms, part 1–3. Berlin, Germany: ISO. ISO 11268-1, 11268–2, 11268–3. 36 p.

Iwai I, Hara T, Sonoda Y. 1975. Factors affecting cadmium uptake by the corn plant. *Soil Sci Plant Nutr* 21:37–46.

Iwasaki K, Takahashi E. 1989. Effects of charge characteristics of Cu-chelates on the Cu uptake from the solution by Italian ryegrass and red clover. *Soil Sci Plant Nutr* 35:145-150.

Iyer VN, Rao KSM, Sarin R. 1989. Speciation of copper, lead and cadmium in natural waters by voltammetric technique. *Ind J Environ Protection* 9:104-112.

Janes N, Playle RC. 1995. Modeling silver binding to gills of rainbow trout (*Onocorynchus mykiss*). *Environ Toxicol Chem* 14:1847–1858.

Janssen M, Bruins A, De Vries TH, Van Straalen NM. 1991. Comparison of cadmium kinetics in four soil arthropod species. *Arch Environ Contam Toxicol* 20:305–312.

Janssen M, Glastra P, Lembrechts J. 1996. Uptake of ^{134}Cs from a sandy soil by two earthworm species: The effect of temperature. *Arch Environ Contam Toxicol* 31:184–191.

Janssen M, Ma W, Van Straalen N. 1993. Biomagnification of metals in terrestrial ecosystems. *Sci Total Environ* (Suppl):511–524.

Janssen RPT, Peijnenburg WJGM, Posthuma R, Van den Hoop MAGT. 1997. Equilibrium partitioning of heavy metals in Dutch field soils. (I) Relationship between metal partition coefficients and soil characteristics. *Environ Toxicol Chem* 16:2470–2478.

Janssen RPT, Posthuma L, Baerselman R, Den Hollander HA, Van Veen RPM, Peijnenburg WJGM. 1997. Equilibrium partitioning of heavy metals in Dutch field soils. (II) Prediction of metal accumulation in earthworms. *Environ Toxicol Chem* 16:2479–2488.

Janssen RPT, Pretorius PJ, Peijnenburg WJGM, Van den Hoop MAGT. 1996. Determination of field-based partition coefficients for heavy metals in Dutch soils and the relationships of these coefficients with soil characteristics. Bilthoven, Netherlands: RIVM. Report nr 719101023. 35 p.

Janzen HH. 1993. Soluble salts. In: Carter MR, editor. Soil sampling and methods of analysis. Canadian Society of Soil Science. Boca Raton FL: Lewis. p 161–166.

Jarvis SC, Jones LHP, Hopper MJ. 1976. Cadmium uptake from solution by plants and its transport from roots to shoots. *Plant Soil* 44:179–191.

Jeng AS, Singh BR. 1993. Partitioning and distribution of cadmium and zinc in selected cultivated soil in Norway. *Soil Sci* 156:240–250.

Jenkinson DS, Ladd JN. 1981. Microbial biomass in soil: Measurement and turnover. In: Paul EA, Ladd JN, editors. Soil biochemistry. Volume 5. New York NY: Marcel Dekker. p 416–471.

Jin X, Bailey GW, Yu YS, Lynch AT. 1996. Kinetics of single and multiple metal ion sorption processes on humic substances. *Soil Sci* 161:509–520.

Jopony M, Young SD 1994. The solidÖsolution equilibria of lead and cadmium in polluted soils. *Eur J Soil Sci* 45:59–70.

Joshi SR, McCrea RC, Shukla BS, Roy JC. 1991. Partitioning and transport of lead-210 in the Ottawa River watershed. *Water Air Soil Pollut* 59:311–320.

Jouanneau JM, Latouche C, Pautrizel F. 1983. Analyse critique des extraction séquentielles à travers l'étude de quelques constituants des résidus d'attaques. *Environ Technol Lett* 4:509–514.

Kalbasi M, Peryea FJ, Lindsay WL, Drake SR. 1995. Measurement of divalent lead activity in lead arsenate contaminated soils. *Soil Sci Soc Am J* 59:1274-1280.

Kalbitz K, Wennrich R. 1998. Mobilization of heavy metals and arsenic in polluted wetland soils and its dependence on dissolved organic matter. *Sci Tot Environ* 209:27–39.

Kennette D, Sauvé S, Hendershot WH. 2001. Uptake of trace metals by the earthworm *Lumbricus terrestris* L. in urban contaminated soils. *Appl Soil Ecol* (in press).

Kerven GL, Edwards DG, Asher CJ. 1984. The determination of native ionic copper concentrations and copper complexation in peat soil extracts. *Soil Sci* 137:91–99.

Kheboian C, Bauer CF. 1987. Accuracy of selective extraction procedures for metal speciation in model aquatic sediments. *Anal Chem* 59:1417–1423.

Killham K, Firestone MK. 1983. Vesicular arbuscular mycorrhizal mediation of grass response to acid and heavy metal deposition. *Plant Soil* 72:39–48.

Kim ND, Fergusson JE. 1991. Effectiveness of a commonly used sequential extraction technique in determining the speciation of cadmium in soils. *Sci Tot Environ* 105:191–210.

Kittrick JA. 1983. Accuracy of several immiscible displacement liquids. *Soil Sci Soc Am J* 47:1045–1047.

Kivalo P, Virtanen R, Wickström K, Wilson M, Pungor E, Horvai G, Tóth K. 1976. An evaluation of some commercial lead(II)-selective electrodes. *Anal Chim Acta* 87:401–409.

Knight B, McGrath SP. 1995. A method to buffer the concentrations of free Zn and Cd ions using a cation exchange resin in bacterial toxicity studies. *Environ Toxicol Chem* 14:2033–2039.

Knight BP, Chaudri AM, McGrath SP, Giller KE. 1998. Determination of chemical availability of cadmium and zinc in soils using inert soil moisture samplers. *Environ Pollut* 99:293–298.

Knight BP, McGrath SP. 1995. A method to buffer the concentrations of free Zn and Cd ions using a cation exchange resin in bacterial toxicity studies. *Environ Toxicol Chem* 14:2033–2039.

Kochian LV. 1993. Zinc absorption from hydroponic solutions by plant roots. In: Robson AD, editor. Zinc in soils and plants. Dordrecht, Netherlands: Kluwer. p 45–57.

Kooijman SALM. 1987. A safety factor for LC50 values allowing for differences in sensitivity among species. *Water Res* 21:269–276.

Kramer JR, Brassard P, Collins P, Clair TA, Takats P. 1990. Variability of organic acids in watersheds. In: Perdue EM, Gjessing ET, editors. Organic acids in aquatic ecosystems. New York NY: Wiley-Interscience. p 127–139.

Krishnamurti GSR, Huang PM, Kozak LM, Rostad HPW, Van Rees KCJ. 1997. Distribution of cadmium in selected soil profiles of Saskatchewan, Canada: Speciation and availability. *Can J Soil Sci* 77:613–619.

Krishnamurti GSR, Huang PM, Van Rees KCJ, Kozak LM, Rostad HPW. 1995. Speciation of particulate-bound cadmium of soils and its bioavailability. *Analyst* 120:659–665.

Kubiak WW, Wang J. 1989. Anodic-stripping voltammetry of heavy metals in the presence of organic surfactants. *Talanta* 8:821–824.

Kubota J, Welch RM, Van Campen DR. 1992. Partitioning of cadmium, copper, lead, and zinc amongst above-ground parts of seed and grain crops grown in selected locations in the USA. *Environ Geochem Health* 14:91–100.

Lage OM, Soares HMVM, Vasconcelos MTSD, Parente AM, Salema R. 1996. Toxicity effects of copper II on the marine dinoflagellate *Amphidinium carterae*: Influence of metal speciation. *Eur J Phycol* 31:341–348.

Lake DL, Kirk PWW, Lester JN. 1984. Fractionation, characterization and speciation of heavy metals in sewage sludge and sludge-amended soils: A review. *J Environ Qual* 13:175–183.

Lanno R, LeBlanc S, Knight B, Tymowski R, Fitzgerald D. 1998. Application of body residues as a tool in the assessment of soil toxicity. In: Sheppard S, Bembridge J, Holmstrup M, Posthuma L, editors. Advances in earthworm ecotoxicology: Proceedings from the 2nd International Workshop on Earthworm Ecotoxicology; 2–5 April 1997; Amsterdam, Netherlands. Pensacola FL: SETAC. p 41–53.

Laskowski R. 1991. Are the top carnivores endangered by heavy metal biomagnification? *Oikos* 60:387–390.

Laurie SH, Manthey JA. 1994. The chemistry and role of metal ion chelation in plant uptake processes. In: Manthey JA, Crowley DE, Luster DG, editors. Biochemistry of metal micronutrients in the rhizosphere. Boca Raton FL: Lewis. p 165–182.

Laurie SH, Tancock NP, McGrath SP, Sanders JR. 1985. The influence of complex formation on trace element uptake by plants. *J Sci Food Agric* 36:541–542.

Laurie SH, Tancock NP, McGrath SP, Sanders JR. 1991a. Influence of complexation on the uptake by plants of iron, manganese, copper and zinc: (I) Effect of EDTA in a multi-metal and computer simulation study. *J Exp Bot* 42:509–513.

Laurie SH, Tancock NP, McGrath SP, Sanders JR. 1991b. Influence of complexation on the uptake by plants of iron, manganese, copper and zinc: (II) Effect of DTPA in a multi-metal and computer simulation study. *J Exp Bot* 42:515–519.

Lebourg A, Sterckeman T, Ciesielski H, Proix N. 1998. Trace metal speciation in three unbuffered salt solutions used to assess their bioavailability in soil. *J Environ Qual* 27:584–590.

Lebourg A, Sterckeman T, Ciesielski H, Proix N. 1996. Intérêt de différents réactifs d'extraction chimique pour l'évaluation de la biodisponsibilité des métaux en traces de sol. *Agronomie* 16:201–215.

Lee DY, Zheng HC. 1993. Chelating resin membrane method for estimation of soil cadmium phytoavailability. *Commun Soil Sci Plant Anal* 24:685–700.

Lee DY, Zheng HC. 1994. Simultaneous extraction of soil phytoavailable cadmium, copper, and lead by chelating resin membrane. *Plant Soil* 164:19–23.

Lee SZ, Allen HE, Huang CP, Sparks DL, Sanders PF, Peijnenburg WJGM. 1996. Predicting soil-water partition coefficients for cadmium. *Environ Sci Technol* 30:3418–3424.

Levy DB, Barbarick KA, Siemer EG, Sommers LE. 1992. Distribution and partitioning of trace metals in contaminated soils near Leadville, Colorado. *J Environ Qual* 21:185–195.

Lexmond TM. 1980. The effect of soil pH on copper toxicity to forage maize grown under field conditions. *Neth J Agric Sci* 28:164–184.

Lexmond TM, Van der Vorm PDJ. 1981. The effect of soil pH on copper toxicity to hydroponically grown maize. *Neth J Agric Sci* 29:209–230.

Li J. 1997. Fractionation and speciation of trace metals in contaminated urban soils from Montréal, Canada [MSc thesis]. Montreal, Canada: McGill Univ-Macdonald Campus, Department of Natural Resource Sciences. 111 p.

Li U-M, Chaney RL, Schneiter AA. 1994. Effect of soil chloride level on cadmium concentration in sunflower kernels. *Plant Soil* 167:275–280.

Li XL, Georg E, Marschner H. 1991. Acquisition of phosphorus and copper by VA-mycorrhizal hyphae and root to shoot transport in white clover. *Plant Soil* 136:49–57.

Liang CN, Schnoenau JJ. 1995. Development of resin membranes as a sensitive indicator of heavy metal toxicity in soil environment. *Intern J Environ Anal Chem* 59:265–275.

Lighthart B, Baham J, Volk VV. 1983. Microbial respiration and chemical speciation in metal-amended soils. *J Environ Qual* 12:543–548.

Lindsay WL. 1979. Chemical equilibria in soils. New York NY: Wiley-Interscience. 44 p.

Lindsay WL, Norvell WA. 1969a. Development of a DTPA micronutrient soil test. Madison WI: American Society of Agronomy. 84 p.

Lindsay WL, Norvell WA. 1969b. Equilibrium relationships of Zn^{2+}, Fe^{2+}, Ca^{2+}, and H^{+} with EDTA and DTPA in soils. *Soil Sci Soc Am Proc* 33:62–68.

Lindsay WL, Norvell WA. 1978. Development of a DTPA soil test for zinc, iron, manganese, and copper. *Soil Sci Soc Am J* 42:421–428.

Logan EM, Pulford ID, Cook GT, Mackenzie AB. 1997. Complexation of Cu^{2+} and Pb^{2+} by peat and humic acid. *Eur J Soil Sci* 48:685–696.

Logan TJ, Chaney RL. 1983. Utilization of municipal wastewater and sludge on land-metals. In: Page AL, Gleason III TL, Smith Jr JE, Iskander IK, Sommers LE, editors. Proceedings of the 1983 Workshop on utilization of municipal wastewater and sludge on land; Riverside, CA. Riverside CA: Univ California. p 235–323.

Lorenz SE, Hamon RE, Holm PE, Domingues HC, Sequeira EM, Christensen TH, McGrath SP. 1997. Cadmium and zinc in plants and soil solutions from contaminated soils. *Plant Soil* 189:21–31.

Lorenz SE, Hamon RE, McGrath SP. 1994. Differences between soil solutions obtained from rhizosphere and non-rhizosphere soils by water displacement and soil centrifugation. *Eur J Soil Sci* 45:431–438.

Lorenz SE, Hamon RE, McGrath SP, Holm PE, Christensen TH. 1994. Applications of fertilizer cations affect cadmium and zinc concentrations in soil solutions and uptake by plants. *Eur J Soil Sci* 45:159–165.

Lumsdon DG, Evans LJ, Bolton KA. 1995. The influence of pH and chloride on the retention of cadmium, lead, mercury, and zinc by soils. *J Soil Contam* 4:137–150.

Lund W. 1990. Speciation analysis: Why and how? *Fresenius' J Anal Chem* 337:557–564.

Luoma SN. 1995. Prediction of metal toxicity in nature from bioassays: Limitations and research needs. In: Tessier A, Turner DR, editors. Metal speciation and bioavailability in aquatic systems. Chichester, UK: J Wiley. p 609–659.

Ma H, Kim SD, Cha DK, Allen HE. 1999. Effect of kinetics of complexation by humic acid on the toxicity of copper to *Ceriodaphnia dubia*. *Environ Toxicol Chem* 18:828–837.

Ma QY, Lindsay WL. 1993. Measurements of free zinc^{2+} activity in uncontaminated and contaminated soils using chelation. *Soil Sci Soc Am J* 57:963–967.

Ma QY, Lindsay WL. 1995. Estimation of Cd^{2+} and Ni^{2+} activities in soils by chelation. *Geoderma* 68:123–133.

Ma WC. 1982. The influence of soil properties and worm-related factors on the concentration of heavy metals in earthworms. *Pedobiol* 24:109–119.

Ma WC. 1987. Heavy metal accumulation in the mole *Talpa europea* and earthworms as an indicator of metal bioavailability in terrestrial environments. *Bull Environ Contam Toxicol* 39:933–938.

Ma WC, Edelman T, Van Beersum I, Jans T. 1983. Uptake of cadmium, zinc, lead, and copper by earthworms near a zinc smelting complex: Influence of soil pH and organic matter. *Bull Environ Contam Toxicol* 30:424–427.

Ma YB, Uren NC. 1997. The fate and transformation of zinc added to soils. *Aust J Soil Res* 35:727-738.

Ma YB, Uren NC. 1998. Transformations of heavy metals added to soil - application of a new sequential extraction procedure. *Geoderma* 84:157–168.

MacLean AJ, Dekker AJ. 1978. Availability of zinc, copper, and nickel to plants grown in sewage-treated soils. *Can J Soil Sci* 58:381–389.

MacRae RK, Smith DE, Swoboda-Colberg N, Meyer JS, Bergman HL. 1999. Copper binding affinity of rainbow trout (*Oncorhynchus mykiss*) and brook trout (*Salvelinus fontinalis*) gills: Implications for assessing bioavailable metal. *Environ Toxicol Chem* 18:1180–1189.

Maeda M, Ohnishi M, Nakagawa G. 1981. Formation constants for Pb(II)-imonodiacetic acid complexes as determined by potentiometry with glass, Pb-amalgam and lead ion-selective electrodes. *J Inorg Nucl Chem* 43:107–110.

Maier NA, McLaughlin MJ, Heap M, Butt M, Smart MK. 1997. Effect of current season applications of calcitic lime on pH, yield and cadmium concentration of potato (*Solanum tuberosum* L.) tubers. *Nut Cycl Agroecosys* 47:1–12.

Manceau A, Boisset MC, Sarret G, Hazemann JL, Mench M, Cambier P, Prost R. 1996. Direct determination of lead speciation in contaminated soils by EXAFS spectroscopy. *Environ Sci Technol* 30:1540–1552.

Marschner H. 1995. Mineral nutrition of higher plants. London, England: AcademicPr.

Marschner H, Römheld V, Horst WJ, Martin P. 1986. Root-induced changes in the rhizosphere: Importance for the mineral nutrition of plants. *Z Pflanzenernähr Bodenk* 149:441–456.

Marschner H, Treeby M, Römheld V. 1989. Role of root-induced changes in the rhizosphere for iron acquisition in higher plants. *Z Pflanzenernähr Bodenk* 152:197–204.

Martell AE, Motekaitis J, Smith RM. 1988. Structure-stability relationships of metal complexes and metal speciation in environmental aqueous solutions. *Environ Toxicol Chem* 7:417–434.

Martin JM, Nirel P, Thomas AJ. 1987. Sequential extractions techniques: Promises and problems. *Mar Chem* 22:313–341.

McBride, M 1989. Reactions controlling heavy metal solubility in soils. *Adv Soil Sci* 10:1–36.

McBride M. 1994. The environmental chemistry of soils. New York NY: Oxford. 406 p.

McBride M, Sauvé S, Hendershot WH. 1997. Solubility control of Cu, Zn, Cd, and Pb in contaminated soils. *Eur J Soil Sci* 48:337–346.

McCarthy LS, Mackay D. 1993. Enhancing ecotoxicological modeling and assessment, body residues and modes of action. *Environ Sci Technol* 27:1719–1728.

McGrath SP. 1994. Effects of heavy metals from sewage sludge on soil microbes in agricultural ecosystems. In: Ross SM, editor. Toxic metals in soil-plant systems. Chichester, UK: J Wiley. p 247–274.

McGrath SP. 1999. Adverse effects of cadmium on soil microflora and fauna. In: McLaughlin MJ, Singh BR, editors. Cadmium in soils and plants. Dordrecht, Netherlands: Kluwer. p 119-218.

McGrath SP, Brookes PC, Giller KE. 1988. Effects of potentially toxic metals in soil derived from past applications of sewage sludge on nitrogen fixation by *Trifolium repens* L. *Soil Biol Biochem* 20:415–424.

McGrath SP, Knight B, Killham K, Preston S, Paton GI. 1999. Assessment of the toxicity of metals in soils amended with sewage sludge using a chemical speciation technique and a lux-based biosensor. *Environ Toxicol Chem* 18:659–663.

McGrath SP, Loveland PJ. 1992. The soil geochemical atlas of England and Wales. Glasgow, UK: Blackie. 101 p.

McGrath SP, Sanders JR, Laurie SH, Tancock NP. 1986. Experimental determinations and computer predictions of trace metal ion concentrations in dilute complex solutions. *Analyst* 111:459–465.

McLaren RG, Crawford DV. 1973a. Studies on soil copper (I): The fractionation of copper in soils. *J Soil Sci* 24:172–181.

McLaren RG, Crawford DV. 1973b. Studies on soil copper (II): The specific adsorption of copper by soils. *J Soil Sci* 24:443–452.

McLaughlin MJ, Andrew SJ, Smart MK, Smolders E. 1998. Effects of sulfate on cadmium uptake by Swiss chard: (I) Effects on complexation and calcium competition in nutrient solution. *Plant Soil* 202:211–216.

McLaughlin MJ, Maier NA, Correll RL, Smart MK, Sparrow LA, McKay A. 1999. Prediction of cadmium concentrations in potato tubers (*Solanum tuberosum* L.) by pre-plant soil and irrigation water analyses. *Aust J Soil Res* 37:191–207.

McLaughlin MJ, Smolders E, Merckx R. 1998. Soil-root interface: Physicochemical processes. In: Huang PM, editor. Soil chemistry and ecosystem health. Madison WI: Soil Science Society of America. p 233–277.

McLaughlin MJ, Smolders E, Merckx R, Maes A. 1997. Plant uptake of cadmium and zinc in chelator-buffered nutrient solution depends on ligand type. In: Ando T, Fujita K, Mae T, Matsumoto H, Mori S, Sekija J, editors. Plant nutrition for sustainable food production and environment. Dordrecht, Netherlands: Kluwer. p 113–118.

McLaughlin MJ, Stevens DP, Zarcinas BA, Cook N. 1999. Testing soils and plants for heavy metals. *Comm Soil Sci Plant Anal* 31:166–170.

McLaughlin MJ, Tiller KG, Beech TA, Smart MK 1994. Soil salinity causes elevated cadmium concentrations in field-grown potato tubers. *J Environ Qual* 23:1013–1018.

McLaughlin MJ, Tiller KG, Smart MK. 1997. Speciation of cadmium in soil solution of saline/sodic soils and relationship with cadmium concentrations in potato tubers. *Aust J Soil Res* 35:1–17.

McLaughlin MJ, Zarcinas BA, Stevens DP, Cook N. 2000. Soil testing for heavy metals. *Comm Soil Sci Plant Anal* 31:1661–1700

Mench M, Fargues S. 1994. Metal uptake by iron-efficient and inefficient oats. *Plant Soil* 165:227-233.

Mench M, Martin E. 1991. Mobilization of cadmium and other metals from two soils by root exudates of *Zea mays* L., *Nicotiana tabacum* L. and *Nicotiana rustica* L. *Plant Soil* 132:187-196.

Mench M, Morel JL, Guckert A. 1987. Metal binding properties of high molecular weight soluble exudates from maize (*Zea mays* L.) roots. *Biol Fertil Soils* 3:165–170.

Mench M, Morel JL, Guckert A, Guillet B. 1988. Metal binding with root exudates of low molecular weight. *Soil Sci* 39:521–527.

Mench M, Vangronsveld J, Didier V, Clijsters H. 1994. Evaluation of metal mobility, plant availability and immobilization by chemical agents in a limed-silty soil. *Environ Pollut* 86:279–286.

Merckx R, Van Ginkel HJ, SinnaeveJ, Cremers A. 1986. Plant-induced changes in the rhizosphere of maize and wheat: (II) Complexation of cobalt, zinc and manganese in the rhizosphere of maize and wheat. *Plant Soil* 96:95–107.

Minnich MM, McBride MB. 1986. Effect of copper activity on carbon and nitrogen mineralization in field-aged copper-enriched soils. *Plant Soil* 91:231–240.

Minnich MM, McBride MB. 1987. Copper activity in soil solution I. Measurement by ion-selective electrode and Donnan dialysis. *Soil Sci Soc Am J* 51:568–572.

Minnich MM, McBride MB, Chaney RL. 1987. Copper activity in soil solution (II): Relation to copper accumulation in young snapbeans. *Soil Sci Soc Am J* 51:573–578.

Morel FMM. 1983. Principles of aquatic chemistry. New York NY: Wiley-Interscience. 446 p.

Morel JL, Mench M, Guckert A. 1986. Measurement of Pb^{2+}, Cu^{2+} and Cd^{2+} binding with mucilage exudates from maize (*Zea mays* L.) roots. *Biol Fertil Soils* 2:29–34.

Morgan AJ, Winter C, Yarwood A, Wilkinson, N. 1995. In-vivo metal substitutions in metal sequestrating subcellular compartments: X-ray mapping in cryosections. *Scanning Micros* 9:1041–1060.

Morgan JE, Morgan AL. 1988. Calcium-lead interactions involving earthworms part 2: The effects of accumulated lead on endogenous calcium in *Lumbricus rubellus*. *Environ Pollut* 55:41–54.

Morris B, Morgan AJ. 1986. Calcium-lead interactions in earthworms: Observations on *Lumbricus terrestris* L. sampled from a calcareous abandoned lead mine site. *Bull Environ Contam Toxicol* 37:226–233.

Morrison GM, Batley GE, Florence TM. 1989. Metal speciation and toxicity. *Chem Br* 8:791–795.

Morrison GMP, Florence TM. 1989. Comparison of physicochemical speciation procedures with metal toxicity to *Chlorella pyrenoidosa*. Copper complexation capacity. *Electroanalysis* 1:107-112.

Mullins GL, Sommers LE. 1986. Cadmium and zinc influx characteristics by intact corn (*Zea mays* L.) seedlings. *Plant Soil* 96:153–164.

Neite H. 1989. Influence of pH and organic carbon content on the solubility of iron, lead, manganese and zinc in forest soils. *Z Pflanzenernaehr Bodenkd* 152:441–446.

Neuhauser EF, Cukic ZV, Malecki MR, Loehr RC, Durkin PR. 1995. Bioconcentration and biokinetics of heavy metals in the earthworm. *Environ Pollut* 89:293–301.

Newman MC, Jagoe CH. 1994. Session 3: Ligands and the bioavailability of metals in aquatic environments. In: Hamelink JL, Landrum PF, Bergman HL, Benson WH, editors. Bioavailability: Physical, chemical, and biological interactions. Boca Raton FL: Lewis. p 39–61.

Nirel PMV, Morel FMM. 1990. Pitfalls of sequential extractions. *Water Res* 24:1055–1056.

Nordgren A, Bååth E, Soderstrom B. 1988. Evaluation of soil respiration characteristics to assess heavy metal effects on soil microorganisms using glutamic acid as a substrate. *Soil Biol Biochem* 20:949–954.

Norvell WA, Lindsay WL. 1969a. Reactions of DTPA chelates of iron, zinc, copper, and manganese with soils. *Soil Sci Soc Am Proc* 36:778–783.

Norvell WA, Lindsay WL. 1969b. Reactions of EDTA complexes of Fe, Zn, Mn, and Cu with soils. *Soil Sci Soc Am Proc* 33:86–91.

Norvell WA. 1984. Comparison of chelating agents as extractants for metals in diverse soil materials. *Soil Sci Soc Am J* 48:1285–1292.

Nye PH. 1966. The effect of nutrient intensity and buffering power of a soil, and the absorbing power, size, and root hairs of a root, on nutrient absorption by diffusion. *Plant Soil* 25:81–104.

Nye PH, Marriott FHC. 1969. A theoretical study of the distribution of substances around roots resulting from simultaneous diffusion and mass flow. *Plant Soil* 30:451–472.

Nye PH, Tinker PB. 1977. Solute movement in the soil-root system. In: Studies in ecology. Volume 4. Oxford, UK: Blackwell. 342 p.

O'Connor GA. 1988. Use and misuse of the DTPA soil test. *J Environ Qual* 17:715–718.

O'Day PA, Carroll SA, Waychunas GA. 1998. Rock-water interactions controlling zinc, cadmium, and lead concentrations in surface waters and sediments, U.S. Tri-State district. (1) Molecular identification using X-ray absorption spectroscopy. *Environ Sci Technol* 32:943–955.

[OECD] Organisation for Economic Cooperation and Development. 1984. Guidelines for the testing of chemicals no. 207: Earthworm acute toxicity test. Adopted 4/4/1984. Paris, France: OECD. 9 p.

Oliver DP, Hannam R, Tiller KG, Wilhelm NS, Merry RH, Cozens GD. 1994. The effects of zinc fertilization on cadmium concentration in wheat grain. *J Environ Qual* 23:705–711.

Opydo J. 1989. The determination of Zn, Cd, Pb, and Cu in soil extracts by anodic stripping voltammetry. *Water Air Soil Pollut* 45:43–48.

Pagenkopf GK.1983. Gill surface interaction model for trace metal toxicity to fishes: Role of complexation, pH, and water hardness. *Environ Sci Technol* 17:342–347.

Palmqvist U, Ahlberg E, Lövberg L, Sjöberg S. 1997. In situ voltammetric determination of metal ions in goethite suspensions: Single metal ion systems. *J Colloid Interface Sci* 196:254–266.

Parametrix. 1995. Persistence, bioaccumulation and toxicity of metals and metal compounds. Ottawa, Canada: ICME. 93 p.

Pardo MT. 1997. Influence of electrolyte on cadmium interaction with selected andisols and alfisols. *Soil Sci* 162:733–740.

Parker DR, Aguilera JJ, Thomason DN. 1992. Zinc-phosphorus interactions in two cultivars of tomato (*Lycopersicon esculentum* L.). *Plant Soil* 143:163–177.

Parker DR, Chaney RL, Norvell WA. 1995. Chemical equilibrium models: Applications to plant nutrition research. In: Loeppert RH, Schwab AP, Goldberg S, editors. Chemical equilibrium and reaction models. Madison WI: SSSA. Special Publication nr 42. p 163–200.

Parker DR, Pedler JF. 1997a. Reevaluating the free-ion activity model of trace metal availability to higher plants. *Plant Soil* 196:223–228.

Parker DR, Pedler JF. 1997b. Reevaluating the free-ion activity model of trace metal availability to higher plants. In: Ando T, Fujita K, Mae T, Matsumoto H, Mori S, Sekija J, editors. Plant nutrition for sustainable food production and environment. Dordrecht, Netherlands: Kluwer. p 107–112.

Parker DR, Pedler JF, Thomason DN, Li H. 1998. Alleviation of copper rhizotoxicity by calcium and magnesium at defined free metal-ion activities. *Soil Sci Soc Am J* 62:965–972.

Paton GI, Rattray EAS, Campbell CD, Meussen H, Cresser MS, Glover LA, Killham K. 1997. Use of genetically modified microbial biosensors for soil ecotoxicity testing. In: Pankhurst CS, Doube B, Gupt V, editors. Bioindicators of soil health. Wallingford, UK: CAB International. p 397–418.

Peijnenburg WJGM, Baerselman R, De Groot AC, Jager T, Posthuma L, Van Veen RPM. 1999. Relating environmental availability to bioavailability: Soil-type–dependent metal accumulation in the oligochaete *Eisenia andrei*. *Ecotoxicol Environ Saf* 44:294–310.

Peijnenburg WJGM, Posthuma L, Eijsackers HJP, Allen HE. 1997. A conceptual framework for implementation of bioavailability of metals for environmental management purposes. *Ecotoxicol Environ Saf* 37:163–172.

Peijnenburg WJGM, Posthuma L, Zweers PGPC, Baerselman R, de Groot AC, Van Veen RPM, Jager T. 1999. Prediction of metal bioavailability in Dutch field soils for the oligochaete *Enchytraeus crypticus*. *Ecotoxicol Environ Saf* 43:170–186.

Perämäki P, Itamies J, Karttunen V, Lajunen LHJ, Pulliainen E. 1992. Influence of pH on the accumulation of cadmium and lead in earthworms (*Aporrectodea caliginosa*) under controlled conditions. *Annu Zool Fenn* 29:105–11.

Peters RW, Shem L. 1995. Treatment of soils contaminated with heavy metals. In: Allen HE, Huang CP, Bailey GW, Bowers AR, editors. Metal speciation and contamination of soil. Boca Raton FL: Lewis. p 255–274.

Petersen R. 1982. Influence of copper and zinc on the growth of a freshwater alga, *Scenedesmus quadricauda*: The significance of chemical speciation. *Environ Sci Technol* 16:443–447.

Petruzzelli G, Petronio BM, Gennaro MC, Vanni A, Lubrano L, Leberator A. 1992. Effect of sewage sludge extract on the sorption process of cadmium and nickel by soil. *Environ Technol* 13:1023–1032.

Phinney JT, Bruland KW. 1994. Uptake of lipophilic organic Cu, Cd, and Pb complexes in the coastal diatom *Thalassiosira weissflogii*. *Environ Sci Technol* 28:1781–1790.

Pierzinsky GW, Schwab AP. 1993. Bioavailability of zinc, cadmium, and lead in a metal-contaminated alluvial soil. *J Environ Qual* 22:247–254.

Pinheiro JP, Mota AM, Simões Gonçalves ML. 1994. Complexation study of humic acids with cadmium(II) and lead(II). *Anal Chim Acta* 284:525–537.

Plankey BJ, Patterson HH. 1987. Kinetics of aluminum-fulvic acid complexation in acidic waters. *Environ Sci Technol* 21:595–601.

Playle RC, Dixon DG, Burnison K. 1993a. Copper and cadmium binding to fish gills: Estimates of metal-gill stability constants and modeling of metal accumulation. *Can J Fish Aquat Sci* 51:2678–2687.

Playle RC, Dixon DG, Burnison K. 1993b. Copper and cadmium binding to fish gills: Modification by dissolved organic carbon and synthetic ligands. *Can J Fish Aquat Sci* 51:2667–2677.

Playle RC, Gensemer RW, Dixon DG. 1992. Copper accumulation on gills of fathead minnows: Influence of water hardness, complexation, and pH of the gill microenvironment. *Environ Toxicol Chem* 11:381–391.

Plette ACC, Nederlof MN, Temminghoff EJM, Van Riemsdijk WH. 1999. Bioavailability of heavy metals in terrestrial and aquatic systems: A quantitative approach. *Environ Toxicol Chem* 18:1882–1890.

Posthuma L, Notenboom J, de Groot AC, Peijnenburg WJGM. 1998. Soil acidity as a major determinant of zinc partitioning and zinc uptake in two oligochaete worms (*Eisenia andrei* and *Enchytraeus crypticus)* exposed in contaminated field soils. In: Sheppard S, Bembridge J, Holmstrup M, Posthuma L, editors. Advances in earthworm ecotoxicology: Proceedings from the 2nd International Workshop on Earthworm Ecotoxicology; 2–5 April 1997; Amsterdam, Netherlands. Pensacola FL: SETAC. p 111–127.

Posthuma L, van Gestel CAM, Smit CE, Bakker DJ, Vonk JW. 1998. Validation of toxicity data and risks limits for soils. Bilthoven, Netherlands: National Institute of Public Health and the Environment. Final report nr 607505004. 230 p.

Procopio JR, Viana MDMO, Hernandez LH. 1997. Microcolumn ion-exchange method for kinetic speciation of copper and lead in natural waters. *Environ Sci Technol* 31:3081–3085.

Prüeß A. 1997. Action values for mobile (NH_4NO_3–extractable) trace elements in soils based on the German national standard DIN 19730. In: Prost R, editor. Contaminated soils: 3rd International Conference on the Biogeochemistry of Trace Elements. Paris, France: Institut National de la Recherche Agronomique. p 415–423.

Pungor E. 1996. Ion-selective electrodes: Analogies and conclusions. *Electroanalysis* 8:348–352.

Qian J, Wang ZJ, Shan XQ, Tu Q, Wen B, Chen B. 1996. Evaluation of plant availability of soil trace metals by chemical fractionation and multiple regression analysis. *Environ Pollut* 91:309–315.

Qiang T, Xiao-Quan S, Jin Q, Zhe-Ming N. 1994. Trace metal redistribution during extraction of model soils by acetic acid/sodium acetate. *Anal Chem* 66:3562–3568.

Quigney D. 1998. An investigation of zinc exposure routes and toxicity in the isopod *Porcellio scaber.* Amsterdam, Netherlands: Free Univ.

Radovanovic H, Koelmans AA. 1998. Prediction of in situ trace metal distribution coefficients for suspended solids in natural waters. *Environ Sci Technol* 32:753–759.

Ramos L, Hernandez LM, Gonzalez MJ. 1994. Sequential fractionation of copper, lead, cadmium, and zinc in soils from or near Donana National Park. *J Environ Qual* 23:50–57.

Rand GM. 1995. Fundamentals of aquatic toxicology. Washington DC: Taylor and Francis.

Rapin F, Tessier A, Campbell PGC, Carignan R. 1986. Potential artifacts in the determination of metal partitioning by a sequential extraction procedure. *Environ Sci Technol* 20:836–840.

Reddy MR, Dunn SJ. 1986. Distribution coefficients for nickel and zinc in soils. *Environ Pollut Ser B Chem Phys* 11:303–313.

Reid R, Brookes JD, Tester MA, Smith FA. 1996. The mechanism of zinc uptake in plants. *Planta* 198:39–45.

Renner R. 1997. Rethinking water quality standards for metal toxicity. *Environ Sci Technol* 31:466A-468A.

Richards BK, Steenhuis TS, Peverly JH, McBride M. 1998. Metal mobility at an old, heavily loaded sludge application site. *Environ Pollut* 99:365–377.

Roca J, Pomares F. 1991. Prediction of available heavy metals by six chemical extractants in a sewage sludge-amended soil. *Commun Soil Sci Plant Anal* 22:2119–2136.

Rogers RD, Williams SE. 1986. Vesicular-arbuscular mycorrhiza: Influence on plant uptake of cesium and cobalt. *Soil Biol Biochem* 18:371–376.

Romein CAFM, Luttik R, Sloof W, Canton JH. 1991. Presentation of a general algorithm for effect assessment based on secondary poisoning (II): Terrestrial food chains. Bilthoven, Netherlands: National Institute of Public Health and the Environment. Report nr 679102007.

Römheld V, Marschner H. 1981 Effect of Fe stress on utilization of Fe chelates by efficient and inefficient plant species. *J Plant Nutr* 3:1–4.

Römkens PF, Dolfing J. 1998. Effect of Ca on the solubility and molecular size of DOC and Cu binding in soil solution samples. *Environ Sci Technol* 32:363–369.

Römkens PF, Salomons W. 1993. The non-applicability of the simple Kd approach in modeling trace behavior: A field study. 9th International Conference Heavy Metals in the Environment. Volume 2; September 1993; Toronto, Canada. Edinburgh, UK: CEP Consultants Ltd. p 496-499.

Ross DS, Bartlett RJ. 1990. Effects of extraction methods and sample storage on properties of solutions obtained from forested spodosols. *J Environ Qual* 19:108–113.

Roth, M. 1992. Metals in invertebrate animals in a forest ecosystem. In: Adriano DC, editor. Biogeochemistry of trace metals. Chelsea MI: Lewis. p 299–238.

Ruby MV, Davis A, Kempton JH, Drexler JW, Bergstrom PD. 1992. Lead bioavailability: Dissolution kinetics under simulated gastric conditions. *Environ Sci Technol* 26:1242–1248.

Ruby MV, Davis A, Nicholson A . 1994. In situ formation of lead phosphates in soils as a method to immobilize lead. *Environ Sci Technol* 28:646–654.

Ruby MV, Davis S, Schoof R, Eberle S, Sellstone CM. 1996. Estimation of lead and arsenic bioavailability using a physiologically based extraction system. *Environ Sci Technol* 30:422-430.

Ruby MV, Schoof R, Brattin W, Goldade M, Post G, Harnois M, Mosby DE, Casteel SW, Berti W, Carpenter M, Edwards D, Cragin D, Chappell W. 1999. Advances in evaluating the oral bioavailability of inorganics in soil for use in human health risk assessment. *Environ Sci Technol* 33:3697–3705.

Rutgers M, Breure AM. 1999. Risk assessment, microbial communities and pollution-induced community tolerance. *Human Ecol Risk Assess* 5:661–670.

Rutgers M, van't Verlaat IM, Wind B, Posthuma L, Breure AM. 1998. Rapid method for assessing pollution-induced community tolerance in contaminated soil. *Environ Toxicol Chem* 17:2210-2213.

Saar RA, Weber JH. 1979. Complexation of cadmium (II) with water- and soil-derived fulvic acids: Effect of pH and fulvic acid concentration. *Can J Chem* 57:1263–1268.

Saar RA, Weber JH. 1980a. Comparison of spectrofluorometry and ion-selective electrode potentiometry for determination of complexes between fulvic acid and heavy-metal ions. *Anal Chem* 52:2095–2099.

Saar RA, Weber JH. 1980b. Lead(II) complexation by fulvic acid: How it differs from fulvic acid complexation of copper(II) and cadmium(II). *Geochim Cosmochim Acta* 44:1381.

Sachdev P, Lindsay WL, Deb DL. 1992. Activity measurements of zinc in soils of different pH using EDTA. *Geoderma* 55:247–257.

Salam AK, Helmke PA. 1998. The pH dependence of free ionic activities and total dissolved concentrations of copper and cadmium in soil solution. *Geoderma* 83:281–291.

Salt C, Kloke A. 1985. Application of $CaCl_2$–extraction for assessment of cadmium and zinc mobility in a wastewater-polluted soil. In: L'Hermite P, editor. Processing and use of organic sludge and liquid agricultural wastes. Dordrecht, Netherlands: D Reidel. p 499–505, 548-550.

Salt DE, Prince RC, Pickering IJ, Raskin I. 1995. Mechanisms of cadmium mobility and accumulation in Indian mustard. *Plant Physiol* 109:1427–1433.

Sanders JR. 1982. The effect of pH upon the copper and cupric ion concentration in soil solutions. *J Soil Sci* 33:679–689.

Santa-Maria GE, Cogliatti DH. 1998. The regulation of zinc uptake in wheat plants. *Plant Sci* 137:1–12.

Santillan-Medrano J, Jurinak JJ. 1975. The chemistry of lead and cadmium in soil: Solid phase formation. *Soil Sci Soc Am Proc* 39:851–856.

Sauerbeck DR, Rietz E. 1982. Soil-chemical evaluation of different extractants for heavy metals in soils. In: Davies RD, Hucker G, L'Hermite P, editors. Environmental effects of organic and inorganic contaminants in sewage sludge. Dordrecht, Netherlands: D Reidel. p 147–160.

Sauvé S. 1999. Chemical speciation, solubility and bioavailability of lead, copper and cadmium in contaminated soils [PhD dissertation]. Ithaca NY: Cornell Univ, Department of Crop and Soil Sciences. 174 p.

Sauvé S, Cook N, Hendershot WH, McBride M. 1996. Linking plant tissue concentration and soil copper pools in urban contaminated soils. *Environ Pollut* 94:154–157.

Sauvé S, Dumestre A, McBride MB, Gillett JW, Berthelin J, Hendershot WH. 1999. Nitrification potential in field-collected soils contaminated with Pb and Cu. *Appl Soil Ecol* 12:29–39.

Sauvé S, Dumestre A, McBride MB, Hendershot WH. 1998. Derivation of soil quality criteria using predicted chemical speciation of Pb^{2+} and Cu^{2+}. *Environ Toxicol Chem* 17:1481–1489.

Sauvé S, Hendershot WH, Allen HE. 2000. Solid-solution partitioning of metals in contaminated soils: Dependence on pH and total metal burden. *Environ Sci Technol* 34:1125–1131.

Sauvé S, Martínez CE, McBride MB, Hendershot WH. 2000. Adsorption of Pb^{2+} by pedogenic and synthetic oxides and leaf compost. *Soil Sci Soc Am J* 64:595–599.

Sauvé S, McBride M, Hendershot WH. 1995. Ion-selective electrode measurements of copper (II) activity in contaminated soils. *Arch Environ Contam Toxicol* 29:373–379.

Sauvé S, McBride M, Hendershot WH. 1997. Speciation of lead in contaminated soils. *Environ Pollut* 98:149–155.

Sauvé S, McBride M, Hendershot WH. 1998a. Lead phosphate solubility in water and soil suspensions. *Environ Sci Technol* 32:388–393.

Sauvé S, McBride MB, Hendershot WH. 1998b. Soil solution speciation of lead(II): Effects of organic matter and pH. *Soil Sci Soc Am J* 62:618–621.

Sauvé S, McBride M, Norvell WA, Hendershot WH. 1997. Copper solubility and speciation of in situ contaminated soils: Effects of copper level, pH and organic matter. *Water Air Soil Pollut* 100:133–149.

Sauvé S, Norvell WA, McBride M, Hendershot WH. 2000. Speciation and complexation of cadmium in extracted soil solutions. *Environ Sci Technol* 34:291–296.

Schalscha EB, Morales M, Ahumada I, Schirado T, Pratt PF. 1980. Fractionation of Zn, Cu, Cr, and Ni in wastewaters, solids and in soil. *Agrochimica* 24:361–368.

Schalscha EG, Morales M, Vergara I, Chang AC. 1982. Chemical fractionation of heavy metals in waste-affected soils. *J Water Pollut Control Fed* 54:175-180.

Scharenberg W, Ebeling E. 1996. Distribution of heavy metals in a woodland food web. *Bull Environ Contam Toxicol* 56:389–396.

Schecher WD, McAvoy DE. 1991. MINEQL+: A chemical equilibrium program for personal computers. User's manual version 2.1. Edgewater MD: Environmental Research Software. 128 p.

Scheidegger AM, Lamble GM, Sparks DL. 1997. Spectroscopic evidence for the formation of mixed-cation hydroxide phases upon metal sorption on clays and aluminum oxides. *J Colloid Interface Sci* 186:118–128.

Scheidegger AM, Strawn DG, Lamble GM, Sparks DL. 1998. The kinetics of mixed Ni-Al hydroxide formation on clays and aluminum oxides: A time-resolved XAFS study. *Geochim Cosmochim Acta* 62:2233–2245.

Scott-Fordsmand JJ, Pedersen MB. 1995. Soil quality criteria for selected inorganic compounds. Working Report Nr 48. Copenhagen, Denmark: Danish EPA. 200 p.

Sekerka I, Lechner JF. 1978. Response of copper(II) selective electrode to some complexing agents. *Anal Lett* A11:415–427.

Selck H, Forbes VE, Gorbes TL. 1998. Toxicity and toxicokinetics of cadmium in *Capitella sp.* (I): Relative importance of water and sediment as routes of cadmium uptake. *Mar Ecol Progr Ser* 164:167–178.

Senesi N. 1992. Metal-humic substance complexes in the environment. Molecular and mechanistic aspects by multiple spectroscopic approach. In: Adriano DC, editor. Biogeochemistry of trace metals. Boca Raton FL: Lewis. p 429–496.

Shaw SC, Rorison IH, Baker AJM. 1984. Solubility of heavy metals in lead mine spoil extracts. *Environ Pollut Ser B Chem Phys* 8:23–33.

Sheppard MI, Thibault DH. 1990. Default soil solid/liquid partition coefficients, K_ds, for four major soil types: A compendium. *Health Phys* 59:471-482.

Sheppard MI, Thibault DH, Mitchell JH. 1987. Element leaching and capillary rise in sandy soil cores: Experimental results. *J Environ Qual* 16:273–284.

Sheppard SC. 1991. A field and literature survey, with interpretation, of elemental concentrations in blueberry *Vaccinium angustifolium*. *Can J Bot* 69:63–77.

Sheppard SC, Evenden WG. 1988a. The assumption of linearity in soil and plant concentration ratios: An experimental evaluation. *J Environ Radioact* 7:221–247.

Sheppard SC, Evenden WG. 1988b. Critical compilation and review of plant/soil concentration ratios for uranium, thorium and lead. *J Environ Radioact* 8:255–286.

Sheppard SC, Evenden WG. 1992. Concentration enrichment of sparingly soluble contaminants (uranium, thorium and lead) by erosion and by soil adhesion to plants and skin. *Environ Geochem Health* 14:121–131.

Sheppard SC, Evenden WG. 1996. Soil-to-plant transfer of elements in natural versus agronomic settings. Technical record TR-742, AECL. Pinawa MB, Canada: Whiteshell Laboratories. p 14.

Sheppard SC, Evenden WG, Cornwell TC. 1997. Depuration and uptake kinetics of I, Cs, Mn, Zn, and Cd by the earthworm (*Lumbricus terrestris*) in radiotracer-spiked litter. *Environ Toxicol Chem* 16:2106–2112.

Sheppard SC, Evenden WG, Pollock RJ. 1989. Uptake of natural radionuclides by field and garden crops. *Can J Soil Sci* 69:761–768.

Sheppard SC, Gaudet C, Sheppard MI, Cureton PM, Wong MP. 1992. The development of assessment and remediation guidelines for contaminated soils: A review of the science. *Can J Soil Sci* 72:359–394.

Sheppard SC, Sheppard MI. 1989. Impact of correlations on stochastic estimates of soil contamination and plant uptake. *Health Phys* 57:653–657.

Sheppard SC, Sheppard MI. 1991. Lead in boreal soils and food plants. *Water Air Soil Pollut* 57-58:79–92.

Shuman M. 1985. Fractionation method for soil microelements. *Soil Sci* 140:11–22.

Shuman MS. 1988. Comparison of anodic stripping voltammetry speciation data with empirical model predictions of pCu. In: Kramer JR, Allen HE, editors. Metal speciation: Theory, analysis and application. Chelsea MI: Lewis. p 125–133.

Sieghardt H. 1987. Heavy metal content and nutrient content of plants and soil samples from metalliferous waste dumps in Bleiberg Carinthia Austria (I) Herbaceous plants. *Z Pflanzenernaehr Bodenkd* 150:129–134.

Siepel H. 1995. Are some mites more ecologically exposed to pollution with lead than others? *Exp Progr Ser* 164:167–178.

Sinnaeve J, Smeulders F, Cremers A. 1983. In situ immobilization of heavy metals with tetraethylenepentamine (Tetren) in natural soils and its effect on toxicity and plant growth. *Plant Soil* 70:49–57.

Skinner FA, Jones PCT, Mollison JE. 1952. A comparison of direct and plate-counting techniques for the quantitative estimation of soil micro-organisms. *J Gen Microbiol* 6:261–271.

Skyllberg U. 1995. Solution/soil ratio and release of cations and acidity from spodosol horizons. *Soil Sci Soc Am J* 59:786–795.

Slavek J, Waller P, Pickering WF. 1990. Labile metal content of sediments: Fractionation scheme based on ion exchange resins. *Talanta* 37:397–406.

Smart RB, Stewart EE. 1985. Differential pulse anodic stripping voltammetry of cadmium(II) at a membrane-covered electrode: Measurement in the presence of model organic compounds. *Environ Sci Technol* 19:137–140.

Smit CE, van Gestel CAM. 1996. Comparison of the toxicity of zinc for the springtail *Folsomia candida* in artificially contaminated and polluted field soils. *Appl Soil Ecol* 3:127–136.

Smit E. 1997. Field relevance of the *Folsomia candida* soil toxicity test [D Phil Thesis]. Amsterdam, Netherlands: Free Univ. 157 p.

Smith FE, Arsenault EA. 1996. Microwave-assisted sample preparation in analytical chemistry. *Talanta* 43:1207–1268.

Smith RM, Martell AE. 1989. Critical stability constants. New York NY: Marcel Dekker.

Smolders E, Bissani C, Helmke PA. 1999. Lime reduces cadmium uptake from soil: Why doesn't it work better? In: Wenzel WW, Adriano DC, Alloway B, Doner HE, Keller C, Lepp NW, Mench M, Naidu R, Pierzynski GM, editors. Proceedings of extended abstracts, 5th International Conference on the Biogeochemistry of Trace Elements. Vienna, Austria: International Society for Trace Element Research. p 528–529.

Smolders E, Lambrechts RM, McLaughlin MJ, Tiller KG. 1997. Effect of soil solution chloride on Cd availability to Swiss chard. *J Environ Qual* 27:426–431.

Smolders E, Lambregts RM, McLaughlin MJ, Tiller KG. 1998. Effect of soil solution chloride on cadmium availability to swiss chard. *J Environ Qual* 27:426–431.

Smolders E, McLaughlin MJ. 1996a. Chloride increases cadmium uptake in Swiss chard in a resin-buffered nutrient solution. *J Soil Sci Soc Am* 60:1443–1447.

Smolders E, McLaughlin MJ. 1996b. Effect of Cl and Cd uptake by swiss chard in nutrient solution. *Plant Soil* 179:57–64.

Smolders E, Van den Brande K, Maes A, Merckx R. 1997. The soil solution composition as a predictor for metal availability to spinach: A comparative study using ^{137}Cs, ^{85}Sr, ^{109}Cd, ^{65}Zn, and ^{152}Eu. In: Ando T, Fujita K, Mae T, Matsumoto H, Mori S, Sekija J, editors. Plant nutrition for sustainable food production and environment. Dordrecht, Netherlands: Kluwer. p 487–492.

Soon YK, Abboud S. 1993. Cadmium, chromium, lead, and nickel. In: Carter MR, editor. Soil sampling and soil analysis. Boca Raton FL: Lewis. p 101–108.

Sparks DL. 1984. Ion activities: An historical and theoretical overview. *Soil Sci Soc Am J* 48:514-518.

Sparks DL. 1995. Environmental soil chemistry. San Diego CA: Academic Pr. 267 p.

Sposito G. 1984. The future of an illusion: Ion activities in soil solutions. *Soil Sci Soc Am J* 48:531-536.

Sposito G. 1989. The chemistry of soils. New York NY: Oxford Univ Pr. 267 p.

Sposito G, Holtzclaw KM. 1977. Titration studies on the polynuclear, polyacidic nature of fulvic acid extracted from sewage sludge-soil mixtures. *Soil Sci Soc Am J* 41:330–336.

Sposito G, Mattigod SV. 1980. GEOCHEM: A computer program for the calculation of chemical equilibria in soil solutions and other natural water systems. Riverside CA: Kearney Foundation of Soil Science, Univ California.

Spurgeon DJ, Hopkin SP. 1995. Extrapolation of the laboratory-based OECD earthworm test to metal-contaminated field sites. *Ecotoxicology* 4:190–205.

Spurgeon DJ, Hopkin SP. 1996. Effects of variation of the organic matter content and pH of soils on the availability and toxicity of zinc for the earthworm *E. fetida*. *Pedobiolie* 40:80–96.

Spurgeon DJ, Hopkin SP. 1999. Comparisons of metal accumulation and excretion kinetics in earthworms (*Eisenia fetida*) exposed to contaminated field and laboratory soils. *Appl Soil Ecol* 11:227–243.

Stella R, Ganzerli-Valentini MT, Borroni PA. 1984. The use of cadmium ion-selective electrode and voltammetric techniques in the study of cadmium complexes with inorganic ligands. In: Pungor E, editor. Ion-selective electrodes 4. Amsterdam, Netherlands: Elsevier. p 633-643.

Stumm W . 1992. Chemistry of the solid-water interface. New York NY: Wiley-Interscience. 428 p.

Stumm W, Morgan JJ. 1996. Aquatic chemistry: Chemical equilibria and rates in natural waters. New York NY: Wiley-Interscience. 1000 p.

Sunda WG, Engel DW, Thuotte RM. 1978. Effect of chemical speciation on toxicity of cadmium to grass shrimp, *Palaemonetes pugio*: Importance of free cadmium ion. *Environ Sci Technol* 12:409–413.

Tack FMG, Verloo MG. 1995. Chemical speciation and fractionation in soil and sediment heavy metal analysis: A review. *Intern J Environ Anal Chem* 59:225–238.

Tambasco G, Sauvé S, Cook N, McBride M, Hendershot WH. 2000. Phytoavailability of Cu and Zn to lettuce *Lactuca sativa* in contaminated soils. *Can J Soil Sci* 80:309–317.

Tanton TW, Crowdy SH. 1971. The distribution of lead chelate in the transpiration stream of higher plants. *Pesticide Sci* 2:211–213.

Taylor GJ, Foy CD. 1985. Differential uptake and toxicity of ionic and chelated copper in *Triticum aestivum*. *Can J Bot* 63:1271–1275.

Taylor RW, Ibeabuci IO, Sistani KR, Shuford JW. 1992. Accumulation of some metals by legumes and their extractability from acid mine spoils. *J Environ Qual* 21:176–180.

Temminghoff EJM, Van der Zee SEATM, de Haan FAM. 1998. Effects of dissolved organic matter on the mobility of copper in a contaminated sandy soil. *Eur J Soil Sci* 49:617–628.

Teo BK. 1986. EXAFS: Basic principles and data analysis: Inorganic chemistry. Berlin, Germany: Springer-Verlag. 349 p.

Tessier A, Buffle J, Campbell PGC. 1994. Uptake of trace metals by aquatic organisms. In: Buffle J, De Vitre R, editors. Chemical and biological regulation of aquatic systems. Boca Raton FL: Lewis. p 197–230.

Tessier A, Campbell PGC. 1988. Comments on testing the accuracy of an extraction procedure for determination of partitioning in sediments by a sequential extraction procedure. *Anal Chem* 60:1475–1476.

Tessier A, Campbell PGC, Bisson M. 1979. Sequential extraction procedure for the speciation of particulate trace metals. *Anal Chem* 51:844–851.

Tessier A, Campbell PGC, Bisson M. 1982. Particulate trace metal speciation in stream sediments and relationships with grain size: Implications for geochemical exploration. *J Geochemic Explor* 16:77–104.

Tiffin LO, Brown JC, Krauss RW. 1960. Differential absorption of metal chelate components by plant roots. *Plant Physiol* 35:362–367.

Tiller KG, Nayyar VK, Clayton PM. 1979. Specific and non-specific sorption of cadmium by soil clays as influenced by zinc and calcium. *Aust J Soil Res* 17:17–28.

Tipping E. 1994. WHAM - A chemical equilibrium model and computer code for waters, sediments, and soils incorporating a discrete site/electrostatic model of ion-binding by humic substances. *Comput Geosci* 21:973–1023.

Tipping E, Hetherington NB, Hilton J, Thompson DW, Bowles E, Hamilton-Taylor J. 1985. Artifacts in the use of selective chemical extraction to determine distributions of metals between oxides of manganese and iron. *Anal Chem* 57:1944–1946.

Tipping E, Hurley MA. 1992. A unifying model of cation binding by humic substances. *Geochim Cosmochim Acta* 56:3627–3641.

Tomson M, Allen HE, English C, Lyman WJ, Pignatello J. 2001. Soil-contaminant interactions. In: Lanno RP, editor. Contaminated soils: From soil-chemical interactions to ecosystem management. Proceedings from the Workshop Assessing Contaminated Soils; 23–27 September 1998; Pellston, MI. Pensacola FL: SETAC.

Town RM, Powell HKJ. 1993. Ion-selective electrode potentiometric studies on the complexation of copper (II) by soil-derived humic and fulvic acids. *Anal Chim Acta* 279:221–233.

Treeby M, Marschner M, Römheld V. 1989. Mobilization of iron and other micronutrient cations from a calcareous soil by plant-borne, microbial and synthetic metal chelators. *Plant Soil* 114:217–226.

Trierweiller JF, Lindsay WL. 1969. EDTA-ammonium carbonate soil test for zinc. *Soil Sci Soc Am Proc* 33:49–54.

Trueby P, Raetz T, Pommer M. 1992. Extraction of heavy metals from soils by ion exchangers. *Z Pflanzenernaehr Bodenkd* 155:95–100.

Turner DR. 1995. Problems in trace metal speciation modeling. In: Tessier A, Turner DR, editors. Metal speciation and bioavailability in aquatic systems. Chichester, UK: J Wiley. p 149–203.

Tyler G. 1983. The impact of heavy metal pollution on forests: A case study of Gusum, Sweden. *Ambio* 13:18–24.

Tyler LD, McBride M. 1982. Influence of Ca, pH, and humic acid on Cd uptake. *Plant Soil* 64:259-262.

Ugolini FC, Corti G, Agneli A, Piccardi F. 1996. Mineralogical, physical, and chemical properties of rock fragments in soil. *Soil Sci* 161:521–542.

Umezawa Y, Tasaki S, Fujiwara S. 1981. Dynamic calibration and memory effects for various kinds of ion-selective electrodes. In: Pungor E, editor. Ion-selective electrodes 3. New York NY: Elsevier. p 359–374.

Ure AM, Quevauviller PH, Muntau H, Griepink B. 1993. Speciation of heavy metals in soils and sediments: An account of the improvement and harmonization of extraction techniques undertaken under the auspices of the BCR of the commission of the European Communities. *Intern J Environ Anal Chem* 51:135–151.

[USEPA] U.S. Environmental Protection Agency. 1993. Standards for the use or disposal of sewage sludge. Federal Register 58:9248–9415.

Van der Watt HVH, Sumner ME, Cabrera ML. 1994. Bioavailability of copper, manganese, and zinc in poultry litter. *J Environ Qual* 23:43–49.

Van Gestel CAM, Dirven-van Breemen EM, Baerselman R. 1993. Accumulation and elimination of cadmium, chromium and zinc and effects on growth and reproduction in *Eisenia andrei* (Oligochaeta, Annelida). *Sci Tot Environ* (Suppl):585–597.

Van Grieken R, Van de Velde R, Robberchet H. 1980. Sample contamination from a commercial grinding unit. *Anal Chim Acta* 118:137–143.

Van Straalen NM. 1996. Critical body concentrations: Their use in bioindication. In: van Straalen NM, Krivolutsky DA, editors. Bioindicator systems for soil pollution. Dordrecht, Netherlands: Kluwer. p 5–16.

Van Straalen NM, Denneman CAJ. 1989. Ecotoxicological evaluation of soil quality criteria. *Ecotoxicol Environ Saf* 18:241–251.

Van Straalen NM, Leeuwangh P, Stortelder PBM. 1994. Progressing limits for soil ecotoxicological risk assessment. In: Donker MH, Eijsackers H, Heimbach F, editors. Ecotoxicology of soil organisms. Boca Raton FL: Lewis. p 397–409.

Van Straalen NM, van Meerendonk JN. 1987. Biological half-lives of lead in *Orchella cincta* (L) (Collembola). *Bull Environ Contam Toxicol* 38:213–219.

Van Wensem J, Vegter JJ, van Straalen NM. 1994. Soil quality criteria derived from critical body concentrations of metals in soil invertebrates. *Appl Soil Ecol* 1:185–191.

Vassil AD, Kapulnik Y, Raskin I, Salt DE. 1998. The role of EDTA in lead transport and accumulation by indian mustard. *Plant Physiol* 117:447–453.

Vercauteren K, Blust R. 1996. Bioavailability of dissolved zinc to the common mussel *Mytilus edulis* in complexing environments. *Mar Ecol Progress Ser* 137:123–132.

Verney MS, Turner DR, Whitfield M, Mantoura RFC. 1984. The use of electrochemical techniques to monitor complexation capacity titrations in natural waters. In: Kramer CJM, Duinker JC, editors. Complexation of trace metals in natural waters. The Hague, Netherlands: Martinus Nijhoff/W. Junk. p 33–46.

Verweij W, Glazewski R, de Haan H. 1992. Speciation of copper in relation to its bioavailability. *Chem Spec Bioavailab* 4:43–51.

Vincent JM. 1970. A manual for the practical study of root-nodule bacteria. Oxford, UK: Blackwell. IBP Handbook nr 15. 164 p.

Vlasov YG, Bychkov EA. 1987. Ion-selective chalcogenide glass electrodes. *Ion-Sel Electrode Rev* 9:5–93.

von Wiren N, Marschner H, Romheld V. 1996. Roots of iron-efficient maize also absorb phytosiderophore-chelated zinc. *Plant Physiol* 111:1119–1125.

Vonk JW, Matla YA, van Gestel CAM, Koolhaas-van Hekezen J, Gerritsen AAM, Henzen L. 1996. The influence of soil characteristics on the toxicity of cadmium for *Folsomia candida, Eisenia fetida* and glutamate mineralisation. Delft, Netherlands: TNO-MEP. TNO report MEP-R 96-144.

Vulkan R, Zhao FJ, Barbosa-Jefferson V, Preston S, Paton GI, Tipping E, McGrath SP. 2000. Copper speciation and impacts on bacterial biosensors in the pore water of copper contaminated soils. *Environ Sci Technol* 34:5115–5121.

Vuori E, Väärikoski J, Hartikainen H, Vakkilainen P, Kumpulainen J, Niinivaara K. 1989. Sorption of selenate by Finnish agricultural soils. *Agric Ecosyst Environ* 25:111–118.

Wagemann R. 1980. Cupric ion-selective electrode and inorganic cationic complexes of copper. *J Phys Chem* 84:3433–3436.

Walker CH, Hopkin SP, Sibly RM, Peakall DB. 1996. Principles of ecotoxicology. London, England: Taylor and Francis.

Wallace A, Romney EM, Alexander GV, Soufi SM, Patel PM. 1977. Some interactions in plants among cadmium, other heavy metals, and chelating agents. *Agron J* 69:18–20.

Wallace A, Romney EM, Cha JW, Soufi SM, Chaudhry FM. 1977. Nickel phytotoxicity in relationship to soil pH manipulation and chelating agents. *Commun Soil Sci Plant Anal* 8:757-764.

Waller PA, Pickering WF. 1990. Evaluation of "labile" metal in sediments by anodic stripping voltammetry. *Talanta* 37:981–983.

Waller PA, Pickering WF. 1991. Evaluation of "labile" metal levels in polluted creek sediments, using transfer of metal to cation exchangers and ASV analysis of chemical extracts. *Chem Speciat Bioavailab* 3:47–54.

Waller PA, Pickering WF. 1992. Determination of "labile" phosphate in lake sediments using anion exchange resins: A critical evaluation. *Chem Speciat Bioavailab* 4:59–68.

Wallwork A. 1983. Earthworm biology. London, England: Edward Arnold.

Weeks JM, Rainbow PS. 1993. The relative importance of food and seawater as sources of copper and zinc to talitrid amphipods (Crustacea; Amphipoda; Talitridea). *J Appl Ecol* 30:722–735.

Wei Y-L, Shyu H-M, Joehuang K-L. 1997. Comparison of microwave versus hot-plate digestion for nine real-world river sediments. *J Environ Qual* 26:764–768.

Weigmann G. 1989. Schadstoffbelastung und bodentiere. landschaftsentwicklung und umweltforschung. *Schriftenr Fachbereits landschaftsentwicklung TU Berl* 59:141-161.

Welch RM, Norvell WA. 1999. Mechanisms of cadmium uptake, translocation and deposition in plants In: McLaughlin MJ, Singh BR, editors. Cadmium in soils and plants. Dordrecht, Netherlands: Kluwer. p 125–150.

Wen X, Du Q, Tang H. 1998. Surface complexation model for the heavy metal adsorption on natural sediment. *Environ Sci Technol* 32:870–875.

Wenzel WW, Blum WEH. 1995. Assessment of metal mobility in soil: Methodological problems. In: Allen HE, Huang CP, Bailey GW, Bowers AR, editors. Metal speciation and contamination of soil. Boca Raton FL: Lewis. p 227–236.

Wenzel WW, Sletten RS, Brandstetter A, Wieshammer G, Stingeder G. 1997. Adsorption of trace metals by tension lysimeters: Nylon membrane versus porous cup. *J Environ Qual* 26:1430–1434.

Westall JC, Morel FMM, Hume DN. 1979. Chloride interference in cupric ion selective electrode measurements. *Anal Chem* 51:1792–1798.

Whalley C, Grant A. 1994. Assessment of the phase selectivity of the European Community Bureau of Reference BCR. sequential extraction procedure for metals in sediment. *Anal Chim Acta* 291:287–295.

Wilkinson HF, Loneragan JF, Quirk JP. 1968. The movement of zinc to plant roots. *Soil Sci Soc Am Proc* 32:831–833.

Will ME, Suter GW. 1995. Toxicological benchmarks for screening potential contaminants of concern for effects on soil and litter invertebrates and heterotrophic process. Oak Ridge TN: Oak Ridge National Laboratory. Report ES-ER-TM-126–R2. 151 p.

Williams CH, David DJ. 1977. Some effects of the distribution of cadmium and phosphate in the root zone on the cadmium content of plants. *Aust J Soil Res* 15:59–68.

Winistörfer D. 1995. Speciation of heavy metals in extracted soil solutions by a cation exchange batch equilibrium method *Commun Soil Sci Plant Anal* 26:1073–1093.

Witter E. 1992. Heavy metal concentrations in agricultural soil critical to microorganisms. Staten Naturvårdverket Rapport 4079. Sweden. 46 p.

Workman SM, Lindsay WL. 1990. Estimating divalent cadmium activities measured in arid-zone soils using competitive chelation. *Soil Sci Soc Am J* 54:987–993.

Xian X. 1987. Chemical partitioning of cadmium zinc lead and copper in soils near smelter. *Environ (II) Sci Health (Part A)* 22:527–542.

Xiao-Quan S, Bin C. 1993. Evaluation of sequential extraction for speciation of trace metals in model soil containing natural minerals and humic acid. *Anal Chem* 65:802–807.

Xiu H, Taylor RW, Shuford JW, Tadesse W. 1991. Comparison of extractants for available sludge-borne metals: A residual study. *Water Air Soil Pollut* 57–58:913–922.

Yamada H, Kang Y, Aso T, Uesugi H, Fujimara T, Yonebayashi K. 1998. Chemical forms and stability of selenium in soil. *Soil Sci Plant Nutr* 44:385–391.

Yeoman S, Sterritt RM, Rudd T, Lester JN. 1989. Particle size fractionation and metal distribution in sewage sludges. *Water Air Soil Pollut* 45:27–42.

Yin Y, Impellitteri CA, You SJ, Allen HE. 2002. The importance of organic matter distribution and extract soil: Solution ratio on the desorption of heavy metals from soils. *Sci Tot Environ* 287:107–119.

Yin Y, Lee S-Z, You S-J, Allen H. 2000. Determinants of metal retention to and release from soils. In: Iskandar I, editor. Environmental restoration of metals-contaminated soils. Boca Raton FL: CRC Pr. p 77–91.

Yin Y, Allen HE, Huang CP, Sanders PF. 1997. Adsorption/desorption isotherms of Hg(II) by soil. *Soil Sci* 162:35–45.

Yin Y, Allen HE, Huang CP, Sparks DL, Sanders PF. 1997. Kinetics of mercury(II) adsorption and desorption by soil. *Environ Sci Technol* 31:496–503.

Yin Y, Allen HE, Li Y, Huang CP, Sanders PF. 1996. Adsorption of mercury (II) by soil: Effects of pH, chloride, and organic matter. *J Environ Qual* 25:837–844.

You SJ, Yin Y, Allen HE. 1999. Partitioning of organic matter in soils: Effects of pH and water:soil ratio. *Sci Tot Environ* 227:155–160.

Zevenhuizen LPTM, Dolfing J, Eshuis EJ, Scholter IJ. 1979. Inhibitory effects of copper on bacteria related to the free ion concentration. *Microb Ecol* 5:139–146.

Zhan T. 1986. A laboratory study of the immobilization of cadmium in soils. *Environ Pollut Series B Chem Phys* 12:265–280.

Zhukov AF, Firer AA, Urusov YI, Vishnajakov AV. 1988. Development and physico-chemical investigation of ion-selective solid-state membrane electrodes for lead, copper, and cadmium ions. In: Pungor E, editor. Ion-selective electrodes 5. Oxford, UK: Pergamon Pr. p 651–659.

Zonneveld P. 1997. De relatie tussen blootstellingsroute, cadmiumopname en cadmiumverdeling bij de pissebed *Porcellio scaber*. Amsterdam, Holland: Stageverslag Afdeling Dieroecologie, Vrije Universiteit.

Index

Bioavailability of Metals in Terrestrial Ecosystems: Importance of Partitioning for Bioavailability to Invertebrates, Microbes, and Plants.
Herbert E. Allen, editor. ISBN 1–880611–46–5

C

E

F

M

N

T

U

V

W

X

Z

SETAC

A Professional Society for Environmental Scientists and Engineers and Related Disciplines Concerned with Environmental Quality

The Society of Environmental Toxicology and Chemistry (SETAC), with offices currently in North America and Europe, is a nonprofit, professional society established to provide a forum for individuals and institutions engaged in the study of environmental problems, management and regulation of natural resources, education, research and development, and manufacturing and distribution.

Specific goals of the society are:

- Promote research, education, and training in the environmental sciences.
- Promote the systematic application of all relevant scientific disciplines to the evaluation of chemical hazards.
- Participate in the scientific interpretation of issues concerned with hazard assessment and risk analysis.
- Support the development of ecologically acceptable practices and principles.
- Provide a forum (meetings and publications) for communication among professionals in government, business, academia, and other segments of society involved in the use, protection, and management of our environment.

These goals are pursued through the conduct of numerous activities, which include:

- Hold annual meetings with study and workshop sessions, platform and poster papers, and achievement and merit awards.
- Sponsor a monthly scientific journal, a newsletter, and special technical publications.
- Provide funds for education and training through the SETAC Scholarship/Fellowship Program.
- Organize and sponsor chapters to provide a forum for the presentation of scientific data and for the interchange and study of information about local concerns.
- Provide advice and counsel to technical and nontechnical persons through a number of standing and ad hoc committees.

SETAC membership currently is composed of more than 5,000 individuals from government, academia, business, and public-interest groups with technical backgrounds in chemistry, toxicology, biology, ecology, atmospheric sciences, health sciences, earth sciences, and engineering.

If you have training in these or related disciplines and are engaged in the study, use, or management of environmental resources, SETAC can fulfill your professional affiliation needs.

All members receive a newsletter highlighting environmental topics and SETAC activities, and reduced fees for the Annual Meeting and SETAC special publications.

All members except Students and Senior Active Members receive monthly issues of *Environmental Toxicology and Chemistry (ET&C)*, a peer-reviewed journal of the Society. Student and Senior Active Members may subscribe to the journal. Members may hold office and, with the Emeritus Members, constitute the voting membership.

If you desire further information, contact the appropriate SETAC office.

SETAC North America
1010 North 12th Avenue
Pensacola, Florida 32501-3367 USA
T 850 469 1500   F 850 469 9778
E setac@setac.org

SETAC Europe
Avenue de la Toison d'Or 67
B-1060 Brussels, Belgium
T 32 2 772 72 81   F 32 2 770 53 83
E setac@setaceu.org

www.setac.org

Environmental Quality Through Science®

Other SETAC Books

Avian Effects Assessment: A Framework for Contaminants Studies
Hart, Balluff, Barfknecht, Chapman, Hawkes, Joermann, Leopold, Luttik, editors
2001

Impact of Low-Dose, High-Potency Herbicides on Nontarget and Unintended Plant Species
Ferenc, editor
2001

Risk Management: Ecological Risk-Based Decision Making
Stahl, Bachman, Barton, Clark, deFur, Ells, Pittinger, Slimak, Wentsel, editors
2001

Ecotoxicology of Amphibians and Reptiles
Sparling, Linder, Bishop, editors
2000

Environmental Contaminants and Terrestrial Vertebrates: Effects on Populations, Communities, and Ecosystems
Albers, Heinz, Ohlendorf, editors
2000

Evaluating and Communicating Subsistence Seafood Safety in a Cross-Cultural Context: Lessons Learned from the Exxon Valdez *Oil Spill*
Field, Fall, Nighswander, Peacock, Varanasi, editors
2000

Multiple Stressors in Ecological Risk and Impact Assessment: Approaches to Risk Estimation
Ferenc and Foran, editors
2000

Natural Remediation of Environmental Contaminants: Its Role in Ecological Risk Assessment and Risk Management
Swindoll, Stahl, Ells, editors
2000

Endocrine Disruption in Invertebrates: Endocrinology, Testing and Assessment
DeFur, Crane, Ingersoll, Tattersfield, editors
1999

Linkage of Effects to Tissue Residues: Development of a Comprehensive Database for Aquatic Organisms Exposed to Inorganic and Organic Chemicals
Jarvinen and Ankley, editors
1999

Multiple Stressors in Ecological Risk and Impact Assessment
Foran and Ferenc, editors
1999

Reproductive and Developmental Effects of Contaminants in Oviparous Vertebrates
DiGiulio and Tillitt, editors
1999

Restoration of Lost Human Uses of the Environment
Grayson Cecil, editor
1999

Ecological Risk Assessment: A Meeting of Policy and Science
Peyster and Day, editors
1998

Ecological Risk Assessment Decision-Support System: A Conceptual Design
Reinert, Bartell, Biddinger, editors
1998

Principles and Processes for Evaluating Endocrine Disruption in Wildlife
Kendall, Dickerson, Geisy, Suk, editors
1998

Radiotelemetry Applications for Wildlife Toxicology Field Studies
Brewer and Fagerstone, editors
1998

Sustainable Environmental Management
Barnthouse, Biddinger, Cooper, Fava, Gillett, Holland, Yosie, editors
1998

Uncertainty Analysis in Ecological Risk Assessment
Warren-Hicks and Moore, editors
1998

Chemical Ranking and Scoring: Guidelines for Relative Assessments of Chemicals
Swanson and Socha, editors
1997

Chemically Induced Alterations in Functional Development and Reproduction of Fishes
Rolland, Gilbertson, Peterson, editors
1997

Ecological Risk Assessment for Contaminated Sediments
Ingersoll, Dillon, Biddinger, editors
1997

Life-Cycle Impact Assessment: The State-of-the-Art, 2nd ed.
Barnthouse, Fava, Humphreys, Hunt, Laibson, Moesoen, Owens, Todd, Vigon, Weitz, Young, editors
1997

Public Policy Application of Life-Cycle Assessment
Allen and Consoli, editors
1997

Quantitative Structure-Activity Relationships (QSAR) in Environment Sciences VII
Chen and Schüürmann, editors
1997

Whole Effluent Toxicity Testing: An Evaluation of Methods and Prediction of Receiving System Impacts
Grothe, Dickson, Reed-Judkins, editors
1996

Procedures for Assessing the Environmental Fate and Ecotoxicity of Pesticides
Mark Lynch, editor
1995

The Multi-Media Fate Model: A Vital Tool for Predicting the Fate of Chemicals
Cowan, D. Mackay, Feijtel, Meent, Di Guardo, Davies, N. Mackay, editors
1995

Aquatic Dialogue Group: Pesticide Risk Assessment and Mitigation
Baker, Barefoot, Beasley, Burns, Caulkins, Clark, Feulner, Giesy, Graney, Griggs, Jacoby, Laskowski, Maciorowski, Mihaich, Nelson, Parrish, Siefert, Solomon, van der Schalie, editors
1994

A Conceptual Framework for Life-Cycle Impact Assessment
Fava, Consoli, Denison, Dickson, Mohin, Vigon, editors
1993

Guidelines for Life-Cycle Assessment: A "Code of Practice"
Consoli, Allen, Boustead, Fava, Franklin, Jensen, Oude, Parrish, Perriman, Postlethewaite, Quay, Seguin, Vigon, editors
1993

A Technical Framework for Life-Cycle Assessment
Fava, Denison, Jones, Curran, Vigon, Selke, Barnum, editors
1991

Research Priorities in Environmental Risk Assessment
Fava, Adams, Larson, G. Dickson, K. Dickson, W. Bishop
1987